Covariant Electrodynamics

Faraday with his induction coil
at the Royal Institution
in London

Newton with his prism
in the ante-chapel at
Trinity College, Cambridge

Maxwell with his color wheel
in George Street, Edinburgh

Covariant Electrodynamics
A Concise Guide

John M. Charap
Emeritus Professor of Theoretical Physics, Queen Mary, University of London

The Johns Hopkins University Press
Baltimore

The Johns Hopkins University Press
2715 North Charles Street
Baltimore, Maryland 21218-4363
www.press.jhu.edu

ISBN-13: 978-1-4214-0014-3(hbk. : alk.paper)
ISBN-13: 978-1-4214-0015-0(pbk. : alk.paper)
ISBN-10: 1-4214-0014-6(hbk. : alk.paper)
ISBN-10: 1-4214-0015-4(pbk. : alk.paper)

Library of Congress Control Number: 2010940271

A catalog record for this book is available from the British Library.

Frontispiece: The frontispiece shows three statues. Although Newton did not contribute directly to electrodynamics, I have included him here because of his seminal contributions to the study of light, especially in his *Opticks*. His statue was celebrated by the poet Wordsworth in his autobiographical poem *The Prelude*, in which he writes of his time as a student at St. John's College, Cambridge:

> And from my bedroom, I in moonlight nights
> Could see right opposite, a few yards off
> The Antechapel, where the Statue stood
> Of Newton with his Prism and silent Face.

My inclusion of Faraday here needs no explanation. His statue stands at the foot of a grand staircase at the Royal Institution where he spent most of his working life. It commemorates in particular his experiments on induction, for which he used the ring he holds in his left hand. The coil, made by winding two coils on opposite sides of the ring with wire insulated by cotton, is still preserved at the Royal Institution.

And of course Maxwell, here a statue unveiled in November 2008. In his hand is a color wheel, an invention he used to demonstrate that any color can be rendered by addition of three primary colors—red, green and blue—not only with light, but also with pigments. This discovery has its application in color printing and in the representation of color on computer and television screens. On the plinth there are reliefs representing his predecessor Newton on one side and his successor Einstein on the other. Another panel is planned to be added to the plinth—it will carry a rendering of Maxwell's equations.

Contents

Preface   ix

1 Introduction   1

2 Mathematical Preliminaries   11
   2.1 A Reminder of Vector Calculus   11
   2.2 Special Relativity 13
   2.3 Four-Vectors 14
   2.4 Covariant and Contravariant Vectors   15
   2.5 Tensors 17
   2.6 Time Dilation and the Lorentz-FitzGerald Contraction   19
   2.7 The Four-Velocity   21
   2.8 Energy and Momentum   22
   2.9 Plane Waves   22
   2.10 Exercises for Chapter 2   22

3 Maxwell's Equations   25
   3.1 Our Starting Point 25
   3.2 The Experimental Background   27
        3.2.1 Coulomb's Law 27
        3.2.2 Absence of Magnetic Monopoles 28
        3.2.3 Ørsted and Ampère 28
        3.2.4 The Law of Biot and Savart 29
        3.2.5 The Displacement Current 29
        3.2.6 Faraday's Law of Induction 30
        3.2.7 The Lorentz Force 30
   3.3 Capacitors and Solenoids 31
        3.3.1 Energy 32
   3.4 Electromagnetic Waves 33
        3.4.1 Polarization 34
        3.4.2 Electromagnetism and Light 34
   3.5 Exercises for Chapter 3 36

4   Behavior under Lorentz Transformations   **37**
   4.1   The Charge-Current Density Four-Vector   37
   4.2   The Lorentz Force   38
   4.3   The Potential Four-Vector   39
   4.4   Gauge Transformations   40
   4.5   The Field-Strength Tensor   43
   4.6   The Dual Field-Strength Tensor   45
   4.7   Exercises for Chapter 4   46

5   Lagrangian and Hamiltonian   **49**
   5.1   Lagrange's Equations   49
   5.2   The Lagrangian for a Charged Particle   51
   5.3   The Hamiltonian for a Charged Particle   53
   5.4   The Lagrangian for the Electromagnetic Field   54
   5.5   The Hamiltonian for the Electromagnetic Field   57
   5.6   Noether's Theorem   58
   5.7   Exercises for Chapter 5   60

6   Stress, Energy, and Momentum   **61**
   6.1   The Canonical Stress Tensor   61
   6.2   The Symmetrical Stress Tensor   63
   6.3   The Conservation Laws with Sources   66
   6.4   The Field as an Ensemble of Oscillators   68
   6.5   Exercises for Chapter 6   69

7   Motion of a Charged Particle   **71**
   7.1   Fields from an Unaccelerated Particle   71
   7.2   Motion of a Particle in an External Field   72
       7.2.1   Uniform Static Magnetic Field   72
       7.2.2   Crossed E and B Fields   73
       7.2.3   Nonuniform Static B-Field   74
       7.2.4   Curved Magnetic Field Lines   75
   7.3   Exercises for Chapter 7   76

8   Fields from Sources   **79**
   8.1   Introducing the Green's Function   79
   8.2   The Delta Function   80
   8.3   The Green's Function   81
   8.4   The Covariant Form for the Green's Function   84
   8.5   Exercises for Chapter 8   85

9   Radiation   **87**
   9.1   Potentials from a Moving Charged Particle   88
   9.2   The Liénard-Wiechert Potentials   89
       9.2.1   Fields from an Unaccelerated Particle   91
       9.2.2   Fields from a Charged Oscillator   94
   9.3   The General Case   96

9.4   The Multipole Expansion   98
      9.4.1   Electric Dipole Radiation 100
      9.4.2   Magnetic Dipole and Higher-Order Terms 102
9.5   Motion in a Circle 103
9.6   Radiation from Linear Accelerators 105
9.7   Radiation from an Antenna 106
9.8   Exercises for Chapter 9 107

10 Media   **109**
10.1 Dispersion 110
      10.1.1 Newton on the "Phænomena of Colours" 111
10.2 Refraction 111
      10.2.1 The Boundary Conditions at the Interface 113
10.3 Čerenkov Radiation 116
10.4 Exercises for Chapter 10 119

11 Scattering   **121**
11.1 Scattering from a Small Scatterer 122
11.2 Many Scatterers 123
11.3 Scattering from the Sky 124
      11.3.1 The Born Approximation 126
      11.3.2 Rayleigh's Explanation for the Blue Sky 128
11.4 Critical Opalescence 133

12 Dispersion   **137**
12.1 The Oscillator Model 137
      12.1.1 The High-Frequency Limit 140
      12.1.2 The Drude Model 140
12.2 Dispersion Relations 142
12.3 The Optical Theorem 145

Epilogue   **149**

Index   **161**

Preface

This book is based on a 30-hour lecture course given for students in the fourth year of the intercollegiate MSci degree at the University of London. It is at a level suitable for final-year undergraduates or beginning research students with theoretical physics aspirations. The topic is classical electromagnetic field theory, which lies at the heart of our understanding of a wide swathe of physical phenomena.

Although I will only rarely make reference to quantum field theory, I have had in mind the application of many of the ideas introduced here to the development of the theory of relativistic quantum fields. Quantum electrodynamics (QED) is still the paradigm example of a relativistic quantum field, and the standard model of high-energy particle physics is modeled on that paradigm. Nor can it be neglected in pursuing string theory and other more recent developments. And QED is best understood after securing a firm grasp on its classical counterpart, the classical electrodynamics derived from Maxwell's equations together with the Lorentz force equation.

Although there are many textbooks that treat electromagnetic theory, I found when teaching a course with that title that none matched the material I thought appropriate. Some are excellent as reference material, and Jackson's *Classical Electrodynamics*[1] is unrivaled for that purpose, but it is too advanced and too comprehensive to serve as an undergraduate text, even for the final year of a four-year program.

I have thought it desirable to give a brief overview of the historical background, especially the experimental findings that provide the basis for the towering achievement of Maxwell in bringing about a synthesis that united electricity and magnetism and embraced optics too. Newton had also sought universal laws that needed no distinction between terrestrial and celestial mechanics, and Maxwell stands beside him in laying the foundation for what we call classical physics. One of the characteristic features of modern physics is a drive towards unification, a search for deep connections between apparently disparate phenomena, so that one may argue that Maxwell's electrodynamics is the bedrock on which much of modern physics rests. Maxwell acknowledged that he did not himself perform the experiments from which he derived his theory. He freely acknowledged his indebtedness above all to the experimental genius of Faraday, that remarkable and sympathetic scientist whose achievements in the laboratory

---

[1] John David Jackson, *Classical Electrodynamics* (Wiley, New York, 1975).

are as inspiring as those of Maxwell in his study.

My second chapter gives a review of some of the mathematical tools that will be used in what follows. At the same time, it serves to introduce some of the notation that is employed in this book, especially that relating to the spacetime of special relativity. Of course, Einstein's theory of special realtivity belongs to the century after Faraday and Maxwell, but I trust that the inclusion of an outline of the use of four-vectors will be excused, despite it being anachronistic to put it before the expression of Maxwell's equations, which were so basic to Einstein's inspiration.

And Maxwell's equations are the subject for Chapter 3, written in the compact form using vectors rather than in the way that Maxwell himself wrote them. The experiments that inspired these equations are summarized, but they do not *imply* the validity of the equations. Rather, they guided Maxwell to his formulation of the equations which so wonderfully agree with the phenomena. And what is to my mind even more remarkable than the fact that so wide a range of experiments can be cited as evidence for their validity is the fact that it is only at the very shortest length scales and the very highest frequencies that the theory breaks down as quantum effects need to be taken into account. Just as Newton's mechanics yielded to Einstein's mechanics when confronted with very high velocities—but still survives as provider of much of the language and many of the concepts used in relativistic theory—so also Maxwell's electrodynamics yielded to quantum electrodynamics. But Maxwell's ideas are still there at the core of relativistic quantum field theory.

The consistency of Maxwell's equations with the Lorentz transformation of special relativity is one of the themes of Chapter 4. The other is the introduction of the scalar and vector potentials and the notion of gauge invariance. Once again the story goes back to Maxwell, although it is not often emphasized that it was he who introduced these potentials. They have very significant counterparts in the gauge theories of high-energy particle physics and also in string theory and M-theory.

The development of analytical mechanics initiated by Lagrange at the end of the eighteenth century and its further extension by Hamilton early in the nineteenth century lend themselves to generalizations for the dynamics of field theories. These are of importance in treatments of quantum fields, and I have thought it appropriate to devote Chapter 5 to the Lagrange and Hamiltonian formulations of Maxwell's electrodynamics.

This leads naturally to a discussion in Chapter 6 of the canonical and the symmetrical stress tensors for the electromagnetic field, the conservation laws of energy and momentum, and (with Noether's theorem) the connection between conservation of charge and the symmetry under gauge transformations. Again, these ideas persist in quantum field theory. The fact that the electromagnetic field can be considered as an ensemble of harmonic oscillators is also shown in this chapter. This idea underlies the famous 1927 paper by Dirac,[2] which may be regarded as laying the foundation for quantum field theory.

---

[2] "The Quantum Theory of the Emission and Absorption of Radiation," *Proc. Roy. Soc. Lond.* **A114**, 243–265 (1927).

There is a change of direction in Chapter 7, which is concerned with the discussion of the motion of a charged particle in given electric and magnetic fields. This is of course an area of wide-ranging practical applications, but I have chosen to restrict attention to only a few. Although most of the matter in the universe exists as plasma (ionized gas), the behavior of which is highly susceptible to magnetic fields, I have not ventured further into a discussion of the consequences of electrodynamics for plasma; so there is no magnetohydro-dynamics, no plasma astrophysics.

The motion of charged particles is influenced by the electromagnetic fields through which they move. And the electromagnetic fields are themselves generated by the charge and current densities ascribed to the moving charged particles. In Chapter 8 the focus is on this aspect of the reciprocal interaction between fields and sources, and in particular on the generation of electromagnetic waves by charge in motion. In electrostatics, the electric field associated with a distribution of electric charge may be determined by considering the charge as a distribution of infinitesimal elements and integrating the Coulomb field associated with each of them. The analogous approach to the problem of determining the potentials for the fields associated with a given distribution of charge and current is to use Green's functions, and Chapter 8 shows how this is to be performed.

Chapter 9 carries this discussion further by introducing the Liénard-Wiechert potentials. They are used to describe radiation from a charged particle executing simple harmonic oscillations, and from a particle moving in a circle, with applications to antennas and to the synchrotron radiation from high-energy accelerators and colliders.

Most of the foregoing chapters have been concerned with charged *particles* and the electromagnetic fields, but the next chapters introduce the phenomena associated with continuous media. Radiation in media propagates at a speed dependent on its frequency, and so we have dispersion (which is treated in more detail in Chapter 12), refraction at interfaces, and polarization on reflection or refraction treated in Chapter 10. A particle traversing a medium at a speed greater than the speed of light in the medium emits radiation, and this Čerenkov radiation is treated in a subsection of Chapter 10.

Even electrically neutral matter has charged constituents, and so it too can scatter radiation. Chapter 11 develops some ideas relating to scattering as, for example, Bragg scattering in crystallography. The scattering of sunlight in the atmosphere is responsible for the blue of the sky, and the simple explanation given by Rayleigh is elaborated as an introduction to the theory of critical opalescence of Einstein and Smoluchowski.

Some of the consequences of dispersion are treated in Chapter 10, but it is in Chapter 12 that we describe the Lorentz oscillator model that underlies the explanation for the dispersive property of polarizable media. The relations between dispersion and absorption of radiation, the Kramers-Kronig dispersion relations, are obtained, as is the optical theorem. Although first encountered in the study of the optical properties of media, dispersion relations and the optical theorem are both of much wider generality, and have found important applications in high-energy particle physics.

I have thought it useful to include a few exercises at the end of the chapters to illustrate some of the ideas and methods employed. But I have found that as an aid to learning, exercises at this level are more of a challenge to the setter than to the student!

I have also added a final chapter pointing to the way that the theory of the electromagnetic field is so much more than a triumph of *classical* physics. It has also proved to be deeply influential on the development of *quantum* field theory, not just for quantum electrodynamics, but more generally for the Standard Model of high-energy physics and current developments in string theory.

The course on electromagnetic theory is still offered at Queen Mary, University of London. I am especially indebted to Bill Spence, who taught the course in the years after I handed it over to him until he, in turn, passed it on. He had to my good fortune preserved the TeX files of my notes that would otherwise have been lost into the electromagnetic ether when my computer collapsed. Others of my colleagues here at Queen Mary have been most supportive and to them too I am indebted. I also thank my editor at the Johns Hopkins University Press, Trevor Lipscombe, and his colleagues there who have forgiven my delays and continued to give me encouragement and support.

Covariant Electrodynamics

Chapter 1

Introduction

Although some of the phenomena of electricity and magnetism were known and described long ago—the ancient Greeks were already aware of the properties of amber ($\mathring{\eta}\lambda\epsilon\kappa\tau\rho o\nu$ = ēlektron in Greek) and of magnetic iron ore (magnetite)—it was not until the seventeenth century that they became topics of major scientific interest and advance, culminating in the work of James Clerk Maxwell (1831–1879).

Many scientists who contributed to the foundations of electromagnetism are celebrated in the units used for some of the quantities that are encountered in its study. The *gilbert* (unit for the magnetic potential) is named after William Gilbert (1544–1603)—a physician to Queen Elizabeth I. His *De Magnete* of 1600 described the magnetic field of the Earth, proposed that the earth is a magnet, and so explained the properties of compasses (which had been known since at least the twelfth century). Gilbert found that friction causes a force in many substances other than amber, including glass, sulfur, sealing wax, and precious stones. He called this an *electric* force. He further noted the differences between electricity and magnetism. Electricity was caused by friction, seemingly attracted everything, and was blocked by screens or water, and caused no definite patterns in attracting other objects. In contrast, magnetism was not affected by friction, only attracted certain other materials, was not blocked by screens or water, and also created definite patterns of attraction (eg iron filings). Gilbert also studied electrostatic phenomena.

No unit is named after Benjamin Franklin (1706–1790) who famously flew a kite into a thunder cloud and was able thereby to charge a *Leyden jar* (an efficient capacitor, invented by Pieter van Musschenbroek), from which he concluded that lightning is an electric phenomenon; he also recognized that the two kinds of charge, resinous and vitreous, were additive inverses. That there was a connection between electricity and magnetism was recorded—but not recognized as such—by mariners who saw that their compasses were affected by lightning strikes. Already by 1766 it was found that the electric force obeys an inverse square law. This was argued by Joseph Priestley (1733–1804), who noted that for an electrified hollow metallic vessel there was no charge on the inside of the vessel, nor any force acting inside. This was precisely like the then known behavior of gravity.

But it was Charles Augustin de Coulomb (1736–1806) whose experiments confirmed the inverse square law of electrostatics, and is remembered by his name as the SI unit *coulomb* (C) of electric charge.

It was only with the invention of the voltaic pile by Count Alessandro Volta (1745–1827) that the study of electricity could move on from electrostatics to electric currents and their effects. And we have the *volt* (V) as the SI unit of electric potential difference or "electromotive force" (e.m.f.).

While setting up a demonstration for a lecture, Hans Christian Ørsted (1777–1851) discovered that an electric current has an effect on a compass needle. The *oersted*, the old CGS unit of magnetic induction, is named after him. Within weeks of learning of Ørsted's discovery, André-Marie Ampère (1775–1836) followed this up with an extensive study of the magnetic phenomena associated with an electric current, showing that a solenoidal current-carrying coil behaves like a magnet. His discovery that parallel wires carrying currents in the same (opposite) direction attract (repel) one another provides the formal definition of the *amp* (A), the SI unit of electric current (and from this also comes the formal definition of the coulomb).

The old CGS unit for magnetic induction, the *gauss*, is named for Carl Friedrich Gauss (1777–1855), a very great mathematician, whose name is also given to the divergence law that features in two of Maxwell's equations. He collaborated in research on magnetism with Wilhelm Weber (1804–1891) whose name is used as the SI unit of magnetic flux, the *weber* (Wb). The SI unit for magnetic flux density is the *tesla* (T) named after Nikola Tesla (1856–1943), an outstanding electrical engineer, whose advocacy alongside George Westinghouse of AC (alternating current) for electric power systems prevailed over Thomas Edison's support for DC (direct current).

As well as units drawn from mechanics (*joule, watt,* etc.), other units used in the physics of electromagnetism include the *henry* (H), the SI unit of inductance, named after Joseph Henry (1797–1878). He discovered the phenomenon of self-inductance independently of Michael Faraday (1791–1867), whose name is given to the *farad* (F), the SI unit of capacitance.

In 1812 Faraday was a 21-year-old bookbinder's assistant. He used to read the books that he was binding and attend popular lectures. Prompted by some of these books and lectures, he wrote to Sir Humphrey Davy (1778–1829) at the Royal Institution, expressing interest in modern developments in electricity and magnetism, and hoping to get any sort of work there. Davy took him on, and, to cut a long story short, by 1829 Faraday had succeeded him as Director. Soon after he began experiments that led to his law of electromagnetic induction.[1] His own description of the crucial experiment, read at the Royal Society in 1831, is characteristic in its lucidty and economy:

> Two hundred and three feet of copper wire in one length were passed round a large block of wood; other two hundred and three feet of similar wire were interposed as a spiral between the turns of the first, and metallic contact everywhere prevented by twine. One of

---

[1] *Faraday's Diary*, Vol. I (G Bell and Sons, London, 1932); Aug. 29th, 1831 et seq.

these helices was connected with a galvanometer and the other with a battery of a hundred pairs of plates four inches square, with double coppers and well charged. When the contact was made, there was a sudden and very slight effect at the galvanometer, and there was also a similar slight effect when the contact with the battery was broken. But whilst the voltaic current was continuing to pass through the one helix, no galvanometrical appearances of any effect like induction upon the other helix could be perceived, although the active power of the battery was proved to be great by its heating the whole of its own helix, and by the brilliancy of the discharge when made through charcoal.

Repetition of the experiments with a battery of one hundred and twenty pairs of plates produced no other effects; but it was ascertained, both at this and at the former time, that the slight deflection of the needle occurring at the moment of completing the connection was always in one direction, and that the equally slight deflection produced when the contact was broken was in the other direction; and, also, that these effects occurred when the first helices were used.

The results which I had by this time obtained with magnets led me to believe that the battery current through one wire did, in reality, induce a similar current through the other wire, but that it continued for an instant only, and partook more of the nature of the electrical wave passed through from the shock of a common Leyden jar than of that from a voltaic battery, and, therefore, might magnetize a steel needle although it scarcely affected the galvanometer.

This expectation was confirmed; for on substituting a small hollow helix, formed round a glass tube, for the galvanometer, introducing a steel needle, making contact as before between the battery and the inducing wire, and then removing the needle before the battery contact was broken, it was found magnetized.

When the battery contact was first made, then an unmagnetized needle introduced, and lastly the battery contact broken, the needle was found magnetized to an equal degree apparently with the first; but the poles were of the contrary kinds.

He later found that if he moved a magnet through a loop of wire, or vice versa, an electric current also flowed in the wire. He used this principle to construct the electric dynamo, the first electric power generator. Faraday had the idea of *lines of magnetic force*, filling space around a magnet, analogous to the streamlines of flow in an incompressible fluid. This concept of lines of flux emanating from charged bodies and magnets provides a way to visualize electric and magnetic fields, a mental model that was crucial to the successful development of electromechanical devices which were to dominate the nineteenth century. And the idea that electromagnetic forces extend into the empty space around the conductor was to germinate as *field theory*. Faraday's demonstrations that a changing magnetic field produces an electric field would subsequently become the basis for one of Maxwell's four equations.

Above all, it was Faraday whose extensive studies of electrical and magnetic phenomena were the foundation for the grand synthesis of elctromagentic theory that was achieved by Maxwell :

> The methods are generally those suggested by the processes of reasoning which are found in the researches of Faraday.[...] By the method which I adopt, I hope to render it evident that I am not attempting to establish any physical theory of a science in which I have hardly made a single experiment, and that the limit of my design is to shew how by a strict application of the ideas and methods of Faraday, the connexion of the very different orders of phenomena which he has discovered may be clearly placed before the mathematical mind.[2]

It is remarkable that the inferences drawn by Maxwell from experimental results obtained with the limited precision accessible to nineteenth-century physics have remained essentially unchanged a century and a half later, and are still at the heart of present-day understanding. To be sure, classical electrodyamics has to be refined to yield *quantum electrodynamics*, but QED has Maxwell's theory as its foundation. Electromagnetism is currently understood as part of the electroweak theory, which unifies the weak nuclear and electromagnetic forces. This in turn is part of the *standard model*. The standard model also includes the strong nuclear force between quarks mediated by gluons, and has so far been extremely successful in predicting phenomena up to energies of more than 100 GeV. The most successful part of the standard model, and the most successful theory in history, is, however, quantum electrodynamics. Its prediction of the magnetic moment of the electron, for example, has been proved correct to within a few parts per billion.[3]

The universality of the speed of propagation of electromagnetic waves implied by Maxwell's equations was the provocation for Einstein's challenge to Newtonian mechanics that led him to the special theory of relativity (1905) and a decade later to his general theory. Running through all of these present-day theories is their use of the notion of a *field*.

Although Faraday had used the word field for the magnetic field, it was in Maxwell's papers that the phrase "electromagnetic field" first appears in English (OED), and the concept of the electromagnetic field was first spelled out in full by Maxwell in a series of papers culminating in "A Dynamical Theory of the Electromagnetic Field" published in 1865 in the *Philosophical Transactions of the Royal Society of London*:

> The theory I propose may therefore be called a theory of the *Electromagnetic Field* because it has to do with the space in the neighbour-

---

[2] " On Faraday's Lines of Force," *Trans. Cambridge Phil. Soc.* **X**, part I (1856).

[3] Writing the gyromagnetic ratio g in terms of the anomaly a via $g = 2(1 + a)$, the world average experimental value is $a_{\text{exp}} = 0.0011596521810...$ (with an uncertainty in the last digit), so the precision of the experimental result is about one part in a trillion! The currently accepted theoretical calculation gives $a_{\text{theory}} = 0.001159652188...$ (again with uncertainties in the last digit).

hood of the electromagnetic bodies, and it may be called a *Dynamical* Theory, because it assumes that in that space there is matter in motion, by which the observed electromagnetic phenomena are produced.

Physicists had long been puzzled by action-at-a-distance. What was it that mediated the force between objects not in direct contact with one another? Newton's interest in this is apparent, for example, in this extract from the Minutes of the Royal Society (9 December 1675):

> That [Newton] having laid upon a table a round piece of glass about two inches broad, in a brass ring, so that the glass might be about one-third of an inch from the table, and the air between them inclosed upon all sides after the manner as if he had whelved a little sieve upon the table: and then rubbing the glass briskly, till some little fragments of paper, laid on the table under the glass, began to be attracted and move nimbly to and fro; after he had done rubbing the glass, the papers would continue a pretty while in various motions; sometimes leaping up to the glass and resting there awhile; then leaping down and resting there, and then leaping up and down again; and this sometimes in lines perpendicular to the table sometimes in oblique ones; sometimes also leaping up in one arch and down in another divers times together, without sensible resting between; sometimes skip in a bow from one part of the glass to another, without touching the table; and sometimes hang by a corner and turn often about very nimbly, as if they had been carried about in the midst of a whirlwind; and he otherwise variously moved every paper with a diverse motion. And upon sliding his finger on the upper side of the glass, though neither the glass nor inclosed air below were moved thereby, yet would the papers, as they hung under the glass receive some new motion inclining this or that way, according as he moved his finger.

> The experiment he proposes to be varied with a larger glass placed farther from the table, and to make use of bits of leaf gold instead of papers, steeming that this will succeed much better, so as perhaps to make the gold rise and fall in spiral lines or whirl for a time in the air, without touching the table or glass. Ordered that this experiment be tried the next meeting.

Newton was opposed to the notion of *action-at-a-distance*, and so had difficulty with what might appear as instances of such. These include the electrostatic phenomena described above, and famously also the force of gravity, which he discusses at length in the "General Scholium" that concludes Book III of the *Principia*. "Hitherto we have explained the phenomena of the heavens and of our sea [the tides] by the power of gravity, but have not yet assigned the cause of this power. [...] I have not been able to discover the cause of those properties from phenomena, and I frame no hypotheses."

> That gravity should be innate, inherent, and essential to matter, so
> that one body may act upon another at a distance through a vacuum
> without the mediation of anything else, by and through which their
> action and force may be conveyed from one to another, is to me so
> great an absurdity that I believe no man who has in philosophical
> matters a competent faculty of thinking, can ever fall into it. (Letter
> to Richard Bentley, 25 February 1692)

Faraday and Maxwell too wished to eliminate action-at-a-distance from their
explanations of electromagnetic phenomena; Faraday put emphasis on force as
the central concept to be explained, and his lines of force were its embodiment.
And these lines of force were his way to visualize what were, in all but name, the
electric and magnetic fields that exerted forces on material bodies.[4] Maxwell
at first concocted an elaborate mechanical model to interpret these forces, but
was to abandon it when he became convinced that the mathematical description
could stand without the model as a scaffold. His equations describe what he
could now call a field, the electromagnetic field, an entity in its own right.

Heinrich Hertz's experiments (1886–1888) confirming Maxwell's prediction
that electromagnetic waves exist and propagate with the same velocity as light
are justly celebrated as providing the clinching support for Maxwell's theory.
As he wrote in his *Electric Waves*,[5] "The object of these experiments was to
test the fundamental hypotheses of the Faraday-Maxwell theory, and the result
of the experiments is to confirm the fundamental hypotheses of the theory."

> I have described the present set of experiments [...] without paying
> special regard to any particular theory; and, indeed, the demon-
> strative power of the experiments is independent of any particular
> theory. Nevertheless, it is clear that the experiments amount to so
> many reasons in favour of that theory of electromagnetic phenomena
> which was first developed by Maxwell from Faraday's views. It also
> appears to me that the hypothesis as to the nature of light which
> is connected with that theory now forces itself upon the mind with
> still stronger reason than heretofore. Certainly it is a fascinating
> idea that the processes in air which we have been investigating rep-
> resent to us on a million-fold larger scale the same processes which
> go on in the neighbourhood of a Fresnel mirror or between the glass
> plates used for exhibiting Newton's rings.

---

[4]Maxwell, in his *Treatise* wrote: 'Faraday [...] never considers bodies as exising with
nothing between them but their distance, and acting on one another according to some function of that distance. He conceives all space as a field of force, the lines of force being in
general curved, and those due to any body extending from it on all sides, their direction being
modified by the presence of other bodies. I think he would [...] have said that the field of
space is full of lines of force, whose arrangement depends on that of the bodies in the field,
and that the mechanical and electrical action on each body is determined by the lines which
abut on it.'

[5]Translated by David Jones (Macmillan and Co., London, 1896; also Dover Publications,
New York, 1962).

What then is Maxwell's theory? Hertz's pithy and much-quoted answer is given in the introduction to *Electric Waves*: "Maxwell's theory is Maxwell's system of equations."

In "A Dynamical Theory of the Electromagnetic Field" Maxwell derived the "General equations of the electromagnetic field" not in the form with which we will be concerned later in this book, because the vector notation had not yet been invented. So Maxwell uses p, q, r, the components of the electric current (actually current density) where we would use $\mathbf{j}$. Thus, for every vector equation, Maxwell has three equations for the relevant components, and with six such sets of vector equations and two scalar equations, ends up with twenty quantities between which he has twenty equations. So in 1864 Maxwell's equations involved the quantities he named as

For electromagnetic momentum	$F\,G\,H$
For magnetic intensity	$\alpha\,\beta\,\gamma$
For electromotive force	$P\,Q\,R$
For current due to true conduction	$p\,q\,r$
For electric displacement	$f\,g\,h$
For total current (including variation of displacement)	$p'\,q'\,r'$
For quantity of free electricity	e
For electric potential	Ψ

To these correspond, in modern notation:

$$F\,G\,H \longmapsto \mathbf{A} \quad \text{(the vector potential)}$$
$$\alpha\,\beta\,\gamma \longmapsto \mathbf{H} \quad \text{(the magnetic field)}$$
$$P\,Q\,R \longmapsto \mathbf{E} \quad \text{(the electric field, not to be confused with the scalar e. m. f.)}$$
$$p\,q\,r \longmapsto \mathbf{j} \quad \text{(electric current density)}$$
$$f\,g\,h \longmapsto \mathbf{D} \quad \text{(the displacement field)}$$
$$p'\,q'\,r' \longmapsto \mathbf{j}_{\text{tot}} \quad \text{(total current density, including displacement current)}$$
$$e \longmapsto \rho \quad \text{(electric charge density)}$$
$$\Psi \longmapsto \Phi \quad \text{(the scalar electric potential)}$$

And his equations (with their modern equivalents) were

(A) *Equations of Total Current*

$p' = p + \frac{df}{dt}$, etc. $\mathbf{j}_{tot} = \mathbf{j} + \frac{\partial \mathbf{D}}{\partial t}$

(B) *Equations of Magnetic Force*

$\mu\alpha = \frac{dH}{dy} - \frac{dG}{dz}$, etc. $\mathbf{B} = \nabla \wedge \mathbf{A}$

(this uses one of the constitutive relations, $\mathbf{B} = \mu\mathbf{H}$)

(C) *Equations of Electric Currents*

$\frac{d\gamma}{dy} - \frac{d\beta}{dz} = 4\pi p'$, etc. $\nabla \wedge \mathbf{H} = \mathbf{j}_{tot}$

(the factor 4π is not present in the rationalized system of units that we adopt)

(D) *Equations of Electromotive Force*

$P = \mu(\gamma\frac{dy}{dt} - \beta\frac{dz}{dt}) - \frac{dF}{dt} - \frac{d\psi}{dx}$, etc.

(Here the first terms corresponding to $\mathbf{v} \wedge \mathbf{B}$, represent the force on a (unit) char
in a conductor "arising from the motion of the conductor itself"; so this set of
Maxwell's equations are equivalent to the Lorentz force equation:

$\mathbf{F} = q(\mathbf{E} + \mathbf{v} \wedge \mathbf{B})$,   with $\mathbf{E} = -\frac{\partial \mathbf{A}}{\partial t} - \nabla\Phi$)

(E) *Equations of Electric Elasticity*

$P = kf$, etc. $\mathbf{E} = \mathbf{D}/\epsilon$

(where $\epsilon = k^{-1}$ is the permittivity, which Maxwell recognized need not be isotro
in a medium)

(F) *Equations of Electric Resistance*

$P = -\chi p$, etc. $\mathbf{E} = \mathbf{j}/\sigma$

(correcting a surprising sign error, which Maxwell had himself changed in the
Treatise of 1873; here σ is the electrical conductivity)

(G) *Equation of Free Electricity*

$e + \frac{df}{dx} + \frac{dg}{dy} + \frac{dh}{dz} = 0$ $\nabla \cdot \mathbf{D} = \rho$

(correcting another sign error, remarkable because consistency of equations (C)
and (H) with (A) cannot be valid without correcting it. Maxwell had also put
this right in the *Treatise*)

(H) *Equation of Continuity*

$\frac{de}{dt} + \frac{dp}{dx} + \frac{dq}{dy} + \frac{dr}{dz} = 0$ $\frac{\partial \rho}{\partial t} + \nabla \cdot \mathbf{j} = 0$

The simplification wrought by recasting these equations in vector notation
was made in 1884 by Oliver Heaviside (1850–1925) who published a series of
papers in the *Electrician* elaborated in the three volumes of his *Electromagnetic
Theory* (1893, 1899, 1912, respectively, reprinted in 1951 in a single volume).
Heaviside too asks the question: "What is Maxwell's theory?" and gives:

> The first approximation to the answer is to say, There is Maxwell's
> book as he wrote it; there is his text, and there are his equations;
> together they make his theory. But when we come to examine it
> more closely, we find that this answer is unsatisfactory. To begin
> with, it is sufficient to refer to papers by physicists, written say
> during the twelve years following the first publication of Maxwell's
> treatise, to see that there may be much difference of opinion as
> to what his theory is. It may be, and has been, interpreted by

different men, which is a sign that it is not set forth in a perfectly clear and unmistakeable form. There are many obscurities and some inconsistencies. Speaking for myself, it was only by changing its form of presentation that I was able to see it clearly, and so as to avoid the inconsistencies.

The Irish physicist George FitzGerald (1851–1901) was one of those who recognized the ground-breaking significance of Maxwell's fundamental paper, and later paid tribute to the role of Heaviside in making his equations more accessible:

Maxwell's treatise is cumbered with the debris of his brilliant lines of assault, of his entrenched camps, of his battles. Oliver Heaviside has cleared these away, has opened up a direct route, has made a broad road, and has explored a considerable trace of country.

In the rest of this book, we will follow the "broad road," using the vector notation introduced by Heaviside.

Chapter 2

Mathematical Preliminaries

2.1 A Reminder of Vector Calculus

To formulate Maxwell's equations most succinctly, we will need the language of vector calculus. We will use rectilinear Cartesian coordinates in a *reference frame K*, so that a point P is identified by (x, y, z). The same point P will have coordinates (x', y', z') in a frame K' rotated with respect to K. For example, if the rotation is through an angle θ about the z-axis,

$$
\begin{aligned}
x' &= x \cos \theta - y \sin \theta, \\
y' &= x \sin \theta + y \cos \theta, \\
z' &= z.
\end{aligned}
\tag{2.1}
$$

A vector may be defined as any object with components that transform in the same way as do the coordinates. So a *vector* $\mathbf{a}$ will have *components* (a_x, a_y, a_z) in K and likewise (a'_x, a'_y, a'_z) in K', which for the rotation between K and K' we have considered as an example will be related by

$$
\begin{aligned}
a'_x &= a_x \cos \theta - a_y \sin \theta, \\
a'_y &= a_x \sin \theta + a_y \cos \theta, \\
a'_z &= a_z.
\end{aligned}
\tag{2.2}
$$

The product

$$
\mathbf{a} \cdot \mathbf{b} = a_x b_x + a_y b_y + a_z b_z
\tag{2.3}
$$

is a *scalar*,

$$
a_x b_x + a_y b_y + a_z b_z = a'_x b'_x + a'_y b'_y + a'_z b'_z.
\tag{2.4}
$$

It is called the *scalar product* between the vectors $\mathbf{a}$ and $\mathbf{b}$. Likewise the product $\mathbf{c} = \mathbf{a} \times \mathbf{b}$ is a vector with components

$$
(c_x, c_y, c_z) = (a_y b_z - a_z b_y, \, a_z b_x - a_x b_z, \, a_x b_y - a_y b_x).
\tag{2.5}
$$

It called the *vector product*. There is an important identity involving these products for any three vectors $\mathbf{a}, \mathbf{b}, \mathbf{c}$,

$$\mathbf{a} \times (\mathbf{b} \times \mathbf{c}) = (\mathbf{a} \cdot \mathbf{c})\mathbf{b} - (\mathbf{a} \cdot \mathbf{b})\mathbf{c}. \tag{2.6}$$

There is also the *scalar triple product*

$$(\mathbf{a} \times \mathbf{b}) \cdot \mathbf{c} = (\mathbf{b} \times \mathbf{c}) \cdot \mathbf{a} = (\mathbf{c} \times \mathbf{a}) \cdot \mathbf{b}. \tag{2.7}$$

A *scalar field* $\Phi(x, y, z)$ is a scalar function of the coordinates (and possibly also of time t). This means that the value of the field in the rotated frame is given by

$$\Phi'(x', y', z') = \Phi(x, y, z). \tag{2.8}$$

Similarly, a *vector field* $\mathbf{V}$ is a vector whose components $\mathbf{V} = (V_x, V_y, V_z)$, are each of them, V_x, V_y, and V_z, functions of (x, y, z) (and possibly t), with values in the rotated frame given by

$$V_x'(x', y', z') = V_x(x, y, z) \cos\theta - V_y(x, y, z) \sin\theta, \text{etc.} \tag{2.9}$$

We wil use partial derivatives $\frac{\partial}{\partial x}, \frac{\partial}{\partial y}, \frac{\partial}{\partial z}$, which are components of a vector $\boldsymbol{\nabla}$, called *del* or *nabla*. From a scalar field Φ we may form a vector field given by

$$\operatorname{grad}\Phi = \boldsymbol{\nabla}\Phi = \left(\frac{\partial\Phi}{\partial x}, \frac{\partial\Phi}{\partial y}, \frac{\partial\Phi}{\partial z}\right). \tag{2.10}$$

From a vector field $\mathbf{V}$ we may form the scalar field

$$\operatorname{div}\mathbf{V} = \boldsymbol{\nabla} \cdot \mathbf{V} = \frac{\partial V_x}{\partial x} + \frac{\partial V_y}{\partial y} + \frac{\partial V_z}{\partial z}. \tag{2.11}$$

Also from the vector field $\mathbf{V}$, we may form another vector field

$$\operatorname{curl}\mathbf{V} = \boldsymbol{\nabla} \times \mathbf{V} = \left(\frac{\partial V_z}{\partial y} - \frac{\partial V_y}{\partial z}, \frac{\partial V_x}{\partial z} - \frac{\partial V_z}{\partial x}, \frac{\partial V_y}{\partial x} - \frac{\partial V_x}{\partial y}\right). \tag{2.12}$$

Some identities are

$$\operatorname{div}(\operatorname{curl}\mathbf{V}) = \boldsymbol{\nabla} \cdot (\boldsymbol{\nabla} \times \mathbf{V}) = 0, \tag{2.13}$$

$$\operatorname{curl}(\operatorname{grad}\Phi) = \boldsymbol{\nabla} \times (\boldsymbol{\nabla}\Phi) = 0, \tag{2.14}$$

$$\boldsymbol{\nabla} \cdot (\Phi\mathbf{V}) = (\mathbf{V} \cdot \boldsymbol{\nabla})\Phi + \Phi\boldsymbol{\nabla} \cdot \mathbf{V}, \tag{2.15}$$

$$\boldsymbol{\nabla} \times (\Phi\mathbf{V}) = (\boldsymbol{\nabla}\Phi) \times \mathbf{V} + \Phi(\boldsymbol{\nabla} \times \mathbf{V}), \tag{2.16}$$

$$\boldsymbol{\nabla} \times (\boldsymbol{\nabla} \times \mathbf{V}) = \boldsymbol{\nabla}(\boldsymbol{\nabla} \cdot \mathbf{V}) - \nabla^2\mathbf{V}. \tag{2.17}$$

There are also some important relations involving integrals. Consider first a closed region of space X surrounded by a boundary surface S. Let $\mathbf{n}$ be a unit vector normal to the boundary surface, pointing *outwards* from the interior, da the infinitesimal area element on the boundary, and $\mathbf{A}$ a vector field. Then we have *Gauss's divergence theorem*:

$$\oint_S \mathbf{A} \cdot \mathbf{n}\,da = \int_X \boldsymbol{\nabla} \cdot \mathbf{A}\,d^3x. \tag{2.18}$$

Now consider a different surface S which has a boundary loop C. If $d\mathbf{l}$ is the infinitesimal line element along C, then we have *Stokes's theorem*:

$$\int_S (\boldsymbol{\nabla} \times \mathbf{A}) \cdot \mathbf{n}\, da = \oint_C \mathbf{A} \cdot d\mathbf{l}. \tag{2.19}$$

2.2   Special Relativity

The fact that the speed c of propagation of the electromagnetic waves predicted by Maxwell's equations is a universal constant, independent of the motion of the source or of the detector of the waves, was the basis on which Einstein (1879–1955) built the special theory of relativity (1905). The basic postulate of relativity is that the fundamental laws of physics have the same form no matter in which frame they are expressed. The prediction that the speed c is independent of the frame of reference is incompatible with the Galilean transformations that gave the relations between quantities in one frame and another, based as they were on the notion that *time* is independent of reference frame, and that *simultaneity* was not relative to the reference frame. Einstein had the courage to replace the Galilean transformations by new ones that incorporate the universality of the speed of light *ab initio*.

What is *special* about the 1905 theory is that it restricts its formulation to *inertial frames of reference*.[1] If K is an inertial frame, so also is any frame K' obtained by a fixed rotation of the axes. So the fundamental laws of nature should preserve their form under that change of coordinates which results from a *rotation* of the frame of reference, which is most simply appreciated by expressing them using the notation of *vectors*. For then so long as both sides of an equation are together scalar, or vector, or whatever, as the case may be, the effect of a rotation is the same on both sides, and what was true in one frame remains true in the rotated frame. The equation is *covariant* under rotations.

Since inertial frames may be in uniform rectilinear motion with respect to one another, the laws of nature must also preserve their form under *boosts*, the transformations of coordinates appropriate to passage from one inertial frame (K) to another (K'), where K' is moving uniformly in a straight line with respect to K.

Maxwell's equations are indeed covariant under rotations, since we regard the electric field $\mathbf{E}$ and the magnetic field $\mathbf{B}$ as vectors. We will see that they are also covariant under boosts.

As usual, we may suppose that the frames K and K' coincide at $t = t' = 0$, and consider a flash of light emanating from their common origin at the instant they coincide. Then the (spherical) wave front described in K by

$$x^2 + y^2 + z^2 = c^2 t^2$$

will be described in K' by

$$x'^2 + y'^2 + z'^2 = c^2 t'^2.$$

---

[1] An inertial frame of reference is one in which Newton's first law of motion holds, that is to say, a frame in which the velocity of any particle remains constant unless there is a net force acting on it.

Homogeneity and isotropy of space and homogeneity of time require a linear relationship between (x', y', z', t') and (x, y, z, t). If we have the "standard orientation" of the axes, so that the frames are parallel, with their relative motion along the x-direction, then consistency of

$$x^2 + y^2 + z^2 = c^2 t^2 \iff x'^2 + y'^2 + z'^2 = c^2 t'^2 \tag{2.20}$$

with linearity of the transformation gives

$$
\begin{aligned}
ct' &= \gamma(ct - \beta x), \\
x' &= \gamma(x - \beta ct), \\
y' &= y, \\
z' &= z,
\end{aligned}
\tag{2.21}
$$

where $\gamma = \frac{1}{\sqrt{1-\beta^2}}$, $\beta = |\boldsymbol{\beta}|$, $\boldsymbol{\beta} = \frac{\mathbf{v}}{c}$, and $\mathbf{v}$ is the relative velocity of K' with respect to K. The inverse relations are

$$
\begin{aligned}
ct &= \gamma(ct' + \beta x'), \\
x &= \gamma(x' + \beta ct'), \\
y &= y', \\
z &= z'.
\end{aligned}
\tag{2.22}
$$

For parallel axes, but when $\mathbf{v}$ is not necessarily along the x-direction, one has

$$
\begin{aligned}
ct' &= \gamma(ct - \boldsymbol{\beta} \cdot \mathbf{x}), \\
\mathbf{x}' &= \gamma \mathbf{x}_{\parallel} + \mathbf{x}_{\perp} - \gamma \boldsymbol{\beta} ct,
\end{aligned}
\tag{2.23}
$$

where $\mathbf{x}_{\parallel} = \frac{\mathbf{x} \cdot \boldsymbol{\beta}}{\beta^2} \boldsymbol{\beta}$ is the component of $\mathbf{x}$ parallel to $\mathbf{v}$ and $\mathbf{x}_{\perp} = \mathbf{x} - \mathbf{x}_{\parallel}$ is the component perpendicular to $\mathbf{v}$.

2.3   Four-Vectors

We now introduce the notation

$$
\begin{aligned}
x^0 &= ct, \\
x^1 &= x, \\
x^2 &= y, \\
x^3 &= z,
\end{aligned}
\tag{2.24}
$$

and define the *rapidity* variable ζ by $\gamma = \cosh \zeta$ (so that $\beta = \tanh \zeta$), and then find that the equations for the Lorentz boost transformation of the coordinates (with standard orientation of the axes) can be written as

$$
\begin{aligned}
x'^0 &= x^0 \cosh \zeta - x^1 \sinh \zeta, \\
x'^1 &= -x^0 \sinh \zeta + x^1 \cosh \zeta, \\
x'^2 &= x^2, \\
x'^3 &= x^3,
\end{aligned}
\tag{2.25}
$$

which is formally very similar to the transformation law for rotations (that is, for a rotation through an angle θ about the x^3- or z-axis);

$$
\begin{aligned}
x'^0 &= x^0, \\
x'^1 &= x^1 \cos\theta + x^2 \sin\theta, \\
x'^2 &= -x^1 \sin\theta + x^2 \cos\theta, \\
x'^3 &= x^3.
\end{aligned}
\tag{2.26}
$$

We call any set of four quantities V^μ, $\mu = 0, 1, 2, 3$, which transform in this fashion under boosts and rotations a *four-vector*; thus

$$
\begin{aligned}
V'^0 &= V^0 \cosh\zeta - V^1 \sinh\zeta, \\
V'^1 &= -V^0 \sinh\zeta + V^1 \cosh\zeta, \\
V'^2 &= V^2, \\
V'^3 &= V^3.
\end{aligned}
\tag{2.27}
$$

Note that if V^μ and U^μ are the components of any two four-vectors, the combination

$$V^0 U^0 - V^1 U^1 - V^2 U^2 - V^3 U^3$$

is an *invariant*, that is to say, its numerical value is unchanged when one replaces V by V' and U by U'. We define this combination as the *(Lorentz) scalar product* between the vectors, and write it as

$$V \cdot U = V^0 U^0 - \mathbf{V} \cdot \mathbf{U}. \tag{2.28}$$

In particular

$$dx^0 dx^0 - dx^1 dx^1 - dx^2 dx^2 - dx^3 dx^3 = c^2 dt^2 - |d\mathbf{x}|^2 \tag{2.29}$$

is invariant, and we write it as ds^2 or as $c^2 d\tau^2$.

Two "nearby" *events* in space-time, separated by dx^μ are said to be *spacelike* separated iff ($=$ if and only if) $ds^2 < 0$, or *timelike* separated iff $ds^2 > 0$, or *null* or *lightlike* separated iff $ds^2 = 0$; and these notions are independent of the reference frame. In the same manner, for events separated in time by Δt and space by $\Delta\mathbf{x}$, the invariant $\Delta s^2 = c^2 \Delta t^2 - |\Delta\mathbf{x}|^2$ is positive iff the events are timelike separated, negative iff they are spacelike separated, or zero iff they are null separated.

2.4   Covariant and Contravariant Vectors

The Lorentz transformation rule for a boost can be expressed as a matrix equation:

$$
\begin{pmatrix} V'^0 \\ V'^1 \\ V'^2 \\ V'^3 \end{pmatrix}
=
\begin{pmatrix}
\cosh\zeta & -\sinh\zeta & 0 & 0 \\
-\sinh\zeta & \cosh\zeta & 0 & 0 \\
0 & 0 & 1 & 0 \\
0 & 0 & 0 & 1
\end{pmatrix}
\begin{pmatrix} V^0 \\ V^1 \\ V^2 \\ V^3 \end{pmatrix}.
\tag{2.30}
$$

Likewise for a rotation:

$$\begin{pmatrix} V'^0 \\ V'^1 \\ V'^2 \\ V'^3 \end{pmatrix} = \begin{pmatrix} 1 & 0 & 0 & 0 \\ 0 & \cos\theta & \sin\theta & 0 \\ 0 & -\sin\theta & \cos\theta & 0 \\ 0 & 0 & 0 & 1 \end{pmatrix} \begin{pmatrix} V^0 \\ V^1 \\ V^2 \\ V^3 \end{pmatrix}. \tag{2.31}$$

More succinctly:

$$V'^\mu = \sum_\nu \Lambda^\mu{}_\nu V^\nu, \tag{2.32}$$

where the elements of the transformation matrix are

$$\Lambda^\mu{}_\nu = \frac{\partial x'^\mu}{\partial x^\nu}.$$

A very useful convention, known as the *Einstein summation convention*, is to omit the summation sign in the previous equation and to write simply

$$V'^\mu = \Lambda^\mu{}_\nu V^\nu, \tag{2.33}$$

it being understood that whenever an index is repeated, it should be summed over: and whenever an index is repeated it will *always* be once "up", and once "down". This form of the transformation rule is valid for any Lorentz transformation, be it a boost or a rotation or a combination of such.

Any four-component object that transforms as

$$V^\mu \to V'^\mu = \Lambda^\mu{}_\nu V^\nu \tag{2.34}$$

is a four-vector, and such a four-vector is called a *contravariant* four-vector, and is always written with the index up. This is to distinguish it from another kind of four-vector, which is written with the index down. An example of this kind is given by the gradient of a scalar f. So if f is a scalar function, the set of its partial derivatives with respect to the coordinates

$$\partial_\alpha f \equiv \frac{\partial f}{\partial x^\alpha} \tag{2.35}$$

transforms as some sort of a vector—but not as a contravariant vector. This is clear from consideration of $df = \partial_\alpha f\, dx^\alpha$, which is of course a scalar, and is thus some sort of a scalar product between the vector whose components are dx^α and the gradient with components $\partial_\alpha f$. The transformation law can easily be derived from

$$\frac{\partial f}{\partial x'^\alpha} = \frac{\partial f}{\partial x^\beta} \frac{\partial x^\beta}{\partial x'^\alpha},$$

which is just the chain rule for differentiation. So the new transformation law is

$$U_\alpha \to U'_\alpha = U_\beta \frac{\partial x^\beta}{\partial x'^\alpha},$$

and a vector with this transformation law is called a *covariant* vector. It is also clear that we may write this as

$$U'_\alpha = U_\beta (\Lambda^{-1})^\beta{}_\alpha. \tag{2.36}$$

2.5   Tensors

Either kind of vector is an example of a more general object called a *tensor*. A tensor is something which has a *linear* transformation rule, in this case (for a Lorentz tensor) under the Lorentz group of transformations from one frame to another. The simplest kind of tensor is one for which the transformation says simply "no change", thus

$$S \rightarrow S' = S, \tag{2.37}$$

and this is characteristic of a *scalar*, which may be called a tensor of rank zero. The *rank* of a tensor is the number of indices it carries. So a vector is a tensor of rank 1. (And we need to specify whether those indices are contravariant or covariant.) A *contravariant tensor of rank 2* is then a two-index quantity with both indices up, say $M^{\alpha\beta}$. Since each index ranges over the four possibilities $(0, 1, 2, 3)$, there are $4 \times 4 = 16$ components, and in general $M^{\alpha\beta} \neq M^{\beta\alpha}$. The transformation law for such a tensor is

$$M^{\alpha\beta} \rightarrow M'^{\alpha\beta} = \Lambda^{\alpha}{}_{\gamma} \Lambda^{\beta}{}_{\delta} M^{\gamma\delta}. \tag{2.38}$$

If $M^{\alpha\beta} = M^{\beta\alpha}$, the tensor is said to be *symmetric*, and this symmetry is preserved in going from one frame to another. Similarly if $M^{\alpha\beta} = -M^{\beta\alpha}$, the tensor is said to be *antisymmetric*, or *skew symmetric*, and again this property is independent of the reference frame. A symmetric second-rank tensor has only ten independent components $(= 4 + \frac{4 \cdot 3}{2})$; an antisymmetric second-rank tensor has six independent components. A *covariant* second-rank tensor will be a two-index quantity like $F_{\mu\nu}$ with transformation law

$$F_{\mu\nu} \rightarrow F'_{\mu\nu} = F_{\rho\sigma}(\Lambda^{-1})^{\rho}{}_{\mu}(\Lambda^{-1})^{\sigma}{}_{\nu}. \tag{2.39}$$

Also to be encountered are *mixed* tensors of the second rank, like $D^{\alpha}{}_{\beta}$ with one contravariant and one covariant index, and the corresponding transformation rule

$$D^{\alpha}{}_{\beta} \rightarrow D'^{\alpha}{}_{\beta} = \Lambda^{\alpha}{}_{\gamma} D^{\gamma}{}_{\delta}(\Lambda^{-1})^{\delta}{}_{\beta}. \tag{2.40}$$

Of special interest is the tensor given by

$$\delta^{\alpha}_{\beta} = \begin{cases} 1, & \text{when } \alpha = \beta; \\ 0, & \text{otherwise.} \end{cases} \tag{2.41}$$

If these are the values of its components in the frame K, then in the frame K' they will be

$$\begin{aligned} \delta'^{\alpha}_{\beta} &= \Lambda^{\alpha}{}_{\gamma} \delta^{\gamma}_{\delta}(\Lambda^{-1})^{\delta}{}_{\beta} \\ &= \Lambda^{\alpha}{}_{\gamma}(\Lambda^{-1})^{\gamma}{}_{\beta} \\ &= \delta^{\alpha}_{\beta}, \end{aligned} \tag{2.42}$$

since thought of as a matrix, δ^{α}_{β} is just the unit matrix:

$$\delta^{\alpha}_{\beta} = \begin{pmatrix} 1 & 0 & 0 & 0 \\ 0 & 1 & 0 & 0 \\ 0 & 0 & 1 & 0 \\ 0 & 0 & 0 & 1 \end{pmatrix}.$$

This means that δ^α_β is an *invariant* tensor.

It is useful to think of the components V^μ of a contravariant vector arranged as a column

$$\begin{pmatrix} V^0 \\ V^1 \\ V^2 \\ V^3 \end{pmatrix},$$

and then the transformation law may be given by matrix multiplication rules as

$$V \to V' = \Lambda V. \tag{2.43}$$

Now the Lorentz transformations keep invariant the form

$$U \cdot V = U^0 V^0 - U^1 V^1 - U^2 V^2 - U^3 V^3 = U^\alpha \eta_{\alpha\beta} V^\beta,$$

where we define

$$\eta_{\alpha\beta} = \begin{cases} +1, & \text{when } \alpha = \beta = 0; \\ -1, & \text{when } \alpha = \beta \neq 0; \\ 0, & \text{when } \alpha \neq \beta. \end{cases} \tag{2.44}$$

Thus for any U, V we have

$$U'^\alpha \eta_{\alpha\beta} V'^\beta = U^\alpha \eta_{\alpha\beta} V^\beta,$$

so that

$$\Lambda^\alpha{}_\gamma U^\gamma \eta_{\alpha\beta} \Lambda^\beta{}_\delta V^\delta = U^\gamma \eta_{\gamma\delta} V^\delta$$

for every U, V, which means that the coefficients of each and every Lorentz transformation have to satisfy

$$\Lambda^\alpha{}_\gamma \eta_{\alpha\beta} \Lambda^\beta{}_\delta = \eta_{\gamma\delta}.$$

Written in terms of matrices, this is

$$\Lambda^{\mathrm{T}} \eta \Lambda = \eta, \tag{2.45}$$

where Λ^{T} is the transpose of Λ. If η had been the unit matrix, this would be the condition that the matrix Λ was orthogonal; as it is, the matrix is said to be *pseudo-orthogonal*. A consequence, which we shall need later, is that the determinant of Λ is 1. The condition on the coefficients $\Lambda^\alpha{}_\beta$ also states that $\eta_{\alpha\beta}$ may be regarded as the components of a constant second-rank symmetrical covariant tensor.

Consider now any contravariant vector V^μ, and define V_μ by $V_\mu = \eta_{\mu\nu} V^\nu$, that is,

$$V_0 = V^0, \quad V_1 = -V^1, \quad V_2 = -V^2, \quad V_3 = -V^3. \tag{2.46}$$

It is then easy to see that V_μ transforms as a covariant four-vector. Thus the tensor $\eta_{\mu\nu}$ can be used to "lower" a contravariant index, thereby giving a covariant index. In an exactly similar way, we may define the constant second-rank symmetrical *contravariant* tensor with components $\eta^{\mu\nu}$ by

$$\eta^{\mu\nu} = \begin{cases} +1, & \text{when } \mu = \nu = 0; \\ -1, & \text{when } \mu = \nu \neq 0; \\ 0, & \text{when } \mu \neq \nu. \end{cases} \tag{2.47}$$

This tensor can be used to "raise" a covariant index. Written as matrices, we have

$$\eta^{\mu\nu} = \eta_{\mu\nu} = \begin{pmatrix} 1 & 0 & 0 & 0 \\ 0 & -1 & 0 & 0 \\ 0 & 0 & -1 & 0 \\ 0 & 0 & 0 & -1 \end{pmatrix}. \tag{2.48}$$

Note also that $U \cdot V = U_\alpha V^\alpha$, which suggest arranging the components of a covariant vector as a row matrix $(U_0 \; U_1 \; U_2 \; U_3)$, and then the scalar product between the two vectors is also the matrix product

$$U \cdot V = U^\alpha \eta_{\alpha\beta} V^\beta = U_\beta V^\beta = \begin{pmatrix} U_0 & U_1 & U_2 & U_3 \end{pmatrix} \begin{pmatrix} V^0 \\ V^1 \\ V^2 \\ V^3 \end{pmatrix}. \tag{2.49}$$

Because the tensor $\eta_{\mu\nu}$ also enters into the formula

$$ds^2 = dx^\mu \eta_{\mu\nu} dx^\nu,$$

it is called the *metric tensor*. In special relativity it is constant, and space-time is *flat*. But in general relativity the metric tensor is *not* constant; one has

$$ds^2 = dx^\mu g_{\mu\nu}(x) dx^\nu, \tag{2.50}$$

and the metric tensor $g_{\mu\nu}$ determines the curvature of space-time. There is a tensor $G_{\mu\nu}$ constructed from the metric tensor and its derivatives. This is the *Einstein tensor*, which is then determined through *Einstein's equations*

$$G_{\mu\nu} = \kappa \, T_{\mu\nu} \tag{2.51}$$

in terms of the density of energy and momentum which appear as the components of the stress-energy-momentum tensor $T_{\mu\nu}$. The constant κ equals $8\pi G/c$, where G is the (Newtonian) gravitational constant.

2.6 Time Dilation and the Lorentz-FitzGerald Contraction

If a particle moves with a three-velocity $\mathbf{u}(t)$ with respect to a frame K, then in the time interval dt (as determined in K) it changes its position by $d\mathbf{x} = \mathbf{u}(t) \, dt$;

so the space-time interval traversed is given by

$$ds^2 = c^2 dt^2 - |d\mathbf{x}|^2 = c^2 dt^2 - \mathbf{u}^2 dt^2$$
$$= c^2 dt^2 \left(1 - \frac{u^2}{c^2}\right)$$
$$= c^2(1 - \beta_u^2)dt^2, \tag{2.52}$$

and this is invariant. Consider the instantaneous *rest-frame*, that is, the frame in which the particle is instantanously at rest. In this frame, the time interval corresponding to dt is called $d\tau$, and evidently the space interval is zero, since the particle is at rest. Thus we have

$$d\tau = \sqrt{1 - \beta_u^2}\, dt,$$

or

$$d\tau = \frac{dt}{\gamma_u} \tag{2.53}$$

which gives the relation between the time interval $d\tau$ (the so-called *proper* time) as measured by a clock moving with the particle, and the time interval dt as measured in the frame K in which the instantaneous speed of the particle is $u(t)$. Since $\gamma \geq 1$, we have

$$t_2 - t_1 = \int_{\tau_1}^{\tau_2} \frac{dt}{d\tau}\, d\tau = \int_{\tau_1}^{\tau_2} \gamma\, d\tau \geq \tau_2 - \tau_1, \tag{2.54}$$

so that *moving clocks run slow*. This is the phenomenon of *time dilation*.

Likewise, if we consider a rod of length L lying along the x-axis, as measured in the frame K in which it is at rest, its end points may be taken to be at $x = 0$, $x = L$. Events A and B which occur at the end points at times t for one end and T for the other thus have coordinates

$$
\begin{aligned}
x_A^0 &= ct, \\
x_A^1 &= 0, \\
x_A^2 &= 0, \\
x_A^3 &= 0
\end{aligned}
$$

and

$$
\begin{aligned}
x_B^0 &= cT, \\
x_B^1 &= L, \\
x_B^2 &= 0, \\
x_B^3 &= 0.
\end{aligned}
$$

These *same* events will, in frame K', have coordinates

$$x_A'^0 = \gamma ct,$$
$$x_A'^1 = -\beta\gamma ct,$$
$$x_A'^2 = 0,$$
$$x_A'^3 = 0$$

and

$$x_B'^0 = \gamma(cT - \beta L),$$
$$x_B'^1 = \gamma(L - \beta cT),$$
$$x_B'^2 = 0,$$
$$x_B'^3 = 0.$$

The length of the rod as determined in K' is the distance between *simultaneous* positions of its end points; that is, one must set $x_A'^0 = x_B'^0$, or $\gamma ct = \gamma(cT - \beta L)$, and then the difference $x_B'^1 - x_A'^1$ between the coordinates of the end points is

$$\begin{aligned} L' &= \gamma(L - \beta cT) - (-\beta\gamma ct) \\ &= \gamma L + \beta\gamma(ct - cT) \\ &= \gamma L + \beta\gamma(-\beta L) \\ &= \frac{1}{\gamma}L. \end{aligned} \tag{2.55}$$

And since $\gamma > 1$, this shows that $L' < L$, which is the *Lorentz-FitzGerald contraction*.

2.7 The Four-Velocity

Since dx^μ is a four-vector, and $d\tau = \frac{1}{c}\sqrt{(dx^0)^2 - (d\mathbf{x})^2}$ is a scalar, it follows that

$$\begin{aligned} \frac{dx^\mu}{d\tau} &= \frac{c\,dx^\mu}{\sqrt{(dx^0)^2 - (d\mathbf{x})^2}} \\ &= \frac{dx^\mu}{dt}\frac{1}{\sqrt{1 - \left(\frac{d\mathbf{x}}{dt}\right)^2\frac{1}{c^2}}} \\ &= \frac{dx^\mu}{dt}\gamma_u \end{aligned}$$

is a four-vector, where $\gamma_u = \frac{1}{\sqrt{1-\beta_u^2}}$, $\beta_u = \frac{u}{c}$, $u = |\mathbf{u}|$, and $\mathbf{u} = \frac{d\mathbf{x}}{dt}$. We write U^μ for this four-vector, giving

$$U^0 = \gamma_u\, c,$$
$$\mathbf{U} = \gamma_u\, \mathbf{u}. \tag{2.56}$$

For a particle with a velocity $\mathbf{u}$ (or better to say, three-velocity $\mathbf{u}$), this four-vector $U = \gamma_u(c, \mathbf{u})$ is called its *four-velocity*. Note that $U^2 \equiv U \cdot U = \gamma_u^2(c^2 - \mathbf{u}^2) = c^2$.

2.8   Energy and Momentum

The nonrelativistic definitions $\mathbf{p} = m\mathbf{u}$ of momentum and KE$= \frac{1}{2}mu^2$ of kinetic energy are replaced by

$$\mathbf{p} = m\mathbf{U} = m\gamma_u\mathbf{u} = m_u\mathbf{u} \tag{2.57}$$

together with

$$p^0 = mU^0 = m\gamma_u c = m_u c, \tag{2.58}$$

so that $p^0 c = m_u c^2$, which may be recognized as the total relativistic energy E for a particle of mass m with speed u. Thus $(E/c, \mathbf{p})$ is a four-vector $= m(U^0, \mathbf{U})$, with

$$p^2 = (E/c)^2 - \mathbf{p}^2 = m^2[(U^0)^2 - \mathbf{U}^2] = m^2 U^2 = m^2 c^2,$$

that is,

$$E^2 = \mathbf{p}^2 c^2 + m^2 c^4, \tag{2.59}$$

so that

$$\begin{aligned}
E &= \sqrt{m^2 c^4 + \mathbf{p}^2 c^2} \\
&= mc^2 + \frac{1}{2}\frac{\mathbf{p}^2}{m} + \cdots
\end{aligned}$$

which shows that the total relativistic energy E has an expansion that leads off with the rest-mass contribution mc^2 and has as its next term the nonrelativistic kinetic energy $\frac{1}{2}\mathbf{p}^2/m = \frac{1}{2}mu^2$.

2.9   Plane Waves

Consider the familiar expression $\exp[i(\mathbf{k}\cdot\mathbf{x} - \omega t)]$ for a plane wave, in which the expression $(\mathbf{k}\cdot\mathbf{x} - \omega t)$ is the *phase*. If the wave is an electromagnetic wave in free space, the phase must be a Lorentz scalar, and this requires, since $x^\mu = (ct, \mathbf{x})$ is a four-vector, that $k^\mu = (\omega/c, \mathbf{k})$ is likewise a four-vector. We note in passing that the de Broglie wave associated with a free relativistic particle of four-momentum p^μ has a four-vector k^μ given by $p^\mu = \hbar k^\mu$, which is a covariant equation combining $E = \hbar\omega$ with $\mathbf{p} = \hbar\mathbf{k}$.

2.10   Exercises for Chapter 2

1   Show explicitly that two successive boosts in the same direction are equivalent to a single boost with a velocity of magnitude

$$v = \frac{v_1 + v_2}{1 + (v_1 v_2/c^2)}.$$

This is the parallel velocity addition law.

2 Muons are particles with a rest mass ≈ 100 MeV/c^2. They are produced in nuclear reactions in the upper atmosphere with energies around 6 GeV. They comprise the largest fraction of energetic charged particles in the cosmic radiation reaching the Earth's surface. (The flux of muons arriving at sea level is about $1/\mathrm{cm}^2/\mathrm{min}$) They decay with a half-life of 2.3 microseconds.

(a)   What is the speed of a 6 GeV muon? [The speed of light is $c = 3 \times 10^8$ m/s. Give your answer as βc.]

(b)   In the reference frame of an observer at rest on the surface of the Earth, how far would a muon travel at this speed in 2.3 μs?

(c)   But the altitude at which the muons are produced is around 15 km. Explain how this apparent paradox can be resolved by using the Lorentz-Fitgerald contraction if working in the frame of reference moving with the muon; or by using relativistic time dilation if working in the frame of reference of the observer at rest.

3   Use the relativistic invariance of *phase* to discuss the Fizeau experiments on the velocity of propogation of light in moving liquids. Show that for a liquid flow at a speed v parallel or antiparallel to the path of the light the speed of the light, as observed in the laboratory, is given to first order in v by

$$u = \frac{c}{n(\omega)} \pm v \left(1 - \frac{1}{n^2} + \frac{\omega}{n} \frac{dn(\omega)}{d\omega} \right),$$

where ω is the frequency of the light in the laboratory (which is the same in the liquid and outside it), and $n(\omega)$ is the index of refraction of the liquid. It may be assumed that the light travels with speed $u' = c/n(\omega')$ relative to the moving liquid.

4   Maxwell's equations are

$$\nabla \cdot \mathbf{D} = \rho,$$

$$\nabla \times \mathbf{H} = \mathbf{j} + \frac{\partial \mathbf{D}}{\partial t},$$

$$\nabla \cdot \mathbf{B} = 0,$$

$$\nabla \times \mathbf{E} = -\frac{\partial \mathbf{B}}{\partial t}.$$

Consider a region of space V bounded by a closed surface S, and also let C be a closed contour in space with an open surface S' spanning the contour. Explaining the notation used, derive from the above equations the integral forms:

$$\oint_S \mathbf{D} \cdot d\mathbf{S} = \int_V \rho \, dV,$$

$$\oint_C \mathbf{H} \cdot d\mathbf{l} = \oint_{S'} \left(\mathbf{j} + \frac{\partial \mathbf{D}}{\partial t} \right) \cdot d\mathbf{S}',$$

$$\oint_S \mathbf{B} \cdot d\mathbf{S} = 0,$$

$$\oint_C \mathbf{E} \cdot d\mathbf{l} = -\int_{S'} \frac{\partial \mathbf{B}}{\partial t} \cdot d\mathbf{S}'.$$

Chapter 3

Maxwell's Equations

3.1 Our Starting Point

Our starting point will be Maxwell's equations:

$$
\begin{aligned}
\nabla \cdot \mathbf{D} &= \rho, \\
\nabla \times \mathbf{H} &= \mathbf{j} + \frac{\partial \mathbf{D}}{\partial t}, \\
\nabla \cdot \mathbf{B} &= 0, \\
\nabla \times \mathbf{E} &= -\frac{\partial \mathbf{B}}{\partial t}.
\end{aligned}
\tag{3.1}
$$

The fields $\mathbf{D}$ (*electric displacement field*) and $\mathbf{H}$ (*magnetic field*) are to be regarded as phenomenological, *macroscopic* fields introduced when discussing media (thought of as continuous, rather than atomic).

In the absence of ferroelectrics or ferromagnets, and for weak applied fields, $\mathbf{D}$ is linearly related to $\mathbf{E}$ and $\mathbf{H}$ to $\mathbf{B}$, so that if we further assume *isotropy* of the medium, we have

$$
\mathbf{D} = \epsilon \mathbf{E}, \qquad \mathbf{H} = \frac{1}{\mu} \mathbf{B},
\tag{3.2}
$$

the *constitutive relations* in their *linear, isotropic* form. More generally $\mathbf{D}$ and $\mathbf{H}$ are determined by complicated, possibly history-dependent (hysteresis) relations from $\mathbf{E}$ and $\mathbf{B}$, with ϵ and μ being tensors rather than scalars since $\mathbf{D}$ may not be parallel to $\mathbf{E}$, nor $\mathbf{H}$ to $\mathbf{B}$.

Under common circumstances one may write

$$
\mathbf{D} = \epsilon_0 \mathbf{E} + \mathbf{P} + \cdots ,
\tag{3.3}
$$

$$
\mathbf{H} = \frac{1}{\mu_0} \mathbf{B} - \mathbf{M} + \cdots ,
\tag{3.4}
$$

which relate those macroscopic fields to the more fundamental fields $\mathbf{E}$ (the *electric field*) and $\mathbf{B}$ (the *magnetic induction*). We have introduced the macroscopically averaged quantities $\mathbf{P}$ and $\mathbf{M}$, the densities of electric and magnetic dipole moment. The ellipses represent contributions from higher moments. The source densities ρ and $\mathbf{j}$ are likewise macroscopic averages of the "free" charges and currents in the medium.

Also of importance is the Lorentz equation for the *force* on a free charge q moving with velocity $\mathbf{v}$,

$$\mathbf{F} = q(\mathbf{E} + \mathbf{v} \times \mathbf{B}). \tag{3.5}$$

In a medium this leads to a generalized version of Ohm's law (which in its simple form is $\mathbf{j} = \sigma\mathbf{E}$) in which $\mathbf{j}$ is given in what may be a complicated fashion in terms of $\mathbf{E}, \mathbf{B}$, and ρ. On equating force to the time derivative of momentum, the Lorentz equation provides an operational definition of the electric field $\mathbf{E}$ and the magnetic induction $\mathbf{B}$ from the force acting on a free *test charge*.

Maxwell's equations, the Lorentz equation, and the constitutive relations for $\mathbf{D}, \mathbf{H}$ and $\mathbf{j}$ are the basic equations of classical electrodynamics in their macroscopic form, useful in the presence of media. Their *microscopic* form, as it were "between the atoms," better described as the *vacuum equations*, are much simpler, since now

$$\mathbf{D} = \epsilon_0\mathbf{E}, \qquad \mathbf{H} = \frac{1}{\mu_0}\mathbf{B}, \tag{3.6}$$

where ϵ_0 and μ_0 are respectively the electric constant (permittivity of the vacuum) and the magnetic constant (permeability of the vacuum) with values

$$\begin{aligned} \epsilon_0 &= 8.854187817\cdots \times 10^{-12} \text{ F} \cdot \text{m}^{-1}, \\ \mu_0 &= 4\pi \times 10^{-7} \text{ N} \cdot \text{A}^{-2}. \end{aligned} \tag{3.7}$$

The sources may be thought of as a set of point charges q_i located at the points $\mathbf{x}_i$, so that

$$\begin{aligned} \rho(\mathbf{x}) &= \sum_i q_i \delta^{(3)}(\mathbf{x} - \mathbf{x}_i), \\ \mathbf{j}(\mathbf{x}) &= \sum_i q_i \mathbf{v}_i \delta^{(3)}(\mathbf{x} - \mathbf{x}_i), \end{aligned} \tag{3.8}$$

where the velocity of the ith charge is $\mathbf{v}_i = \dot{\mathbf{x}}_i$ and the delta function will be discussed in more detail later.

So Maxwell's equations in vacuum now simplify to

$$\begin{aligned} \nabla \cdot \mathbf{E} &= \rho/\epsilon_0, \\ \nabla \times \mathbf{B} &= \mu_0\mathbf{j} + \epsilon_0\mu_0\frac{\partial\mathbf{E}}{\partial t}, \\ \nabla \cdot \mathbf{B} &= 0, \\ \nabla \times \mathbf{E} &= -\frac{\partial\mathbf{B}}{\partial t}. \end{aligned} \tag{3.9}$$

3.2   The Experimental Background

3.2.1   Coulomb's Law

The first equation:

$$\nabla \cdot \mathbf{E} = \rho/\epsilon_0,$$

may be derived from *Coulomb's law* of electrostatics. Charles Augustin de Coulomb had developed the use of a torsion balance to investigate small forces. He applied this technique to determine the force between electrically charged bodies, publishing his results in a series of *Mémoires sur l'Electricité et le Magnétisme* in 1785. His conclusions for the force between charges Q and q a distance r apart may be expressed in the familiar form:

$$F = k\frac{Qq}{r^2}, \tag{3.10}$$

where k is a constant.

If we consider a small body carrying a charge Q, since from the Lorentz force equation the field strength E is determined from the force exerted on a *test charge* q, Coulomb's law implies that the field produced by the charge at position $\mathbf{r}$ away from it is

$$\mathbf{E} = k\frac{Q}{r^2}\frac{\mathbf{r}}{r}.$$

This means that if S is any closed surface, with outward-directed unit normal $\mathbf{n}$, the flux of the electric field across a surface element $\mathbf{n}da$ of S at position $\mathbf{r}$ away from Q is $k\frac{Q}{r^2}\frac{\mathbf{n \cdot r}}{r}da$. But writing $d\Omega$ for the solid angle subtended by the surface element at Q, we find that $\frac{1}{r^2}\frac{\mathbf{n \cdot r}}{r}da$ is $\pm d\Omega$, with the plus sign if the vector $\mathbf{r}$ crosses the surface from inside to out and the minus sign if from outside to in. So we may write $\pm kQ\,d\Omega$ for the flux across our surface element. If Q is *outside* the surface S, we have $\oint_S d\Omega = 0$, since the contributions from inside to outside crossings cancel pairwise against those from outside to inside. But if Q is *inside* S, we have $\oint_S d\Omega = 4\pi$. So the total flux of the electric field across the surface S is

$$\oint_S \mathbf{E} \cdot \mathbf{n}\,da = 4\pi kQ$$

if Q is inside S, and vanishes if Q is outside S. Considering a distribution of charge with density ρ as being built up from small elements, we may conclude that

$$\oint_S \mathbf{E} \cdot \mathbf{n}\,da = 4\pi k \int_X \rho\,d^3x,$$

the integration being over the interior of S. But since from Gauss's law

$$\oint_S \mathbf{E} \cdot \mathbf{n}\,da = \int_X \nabla \cdot \mathbf{E}\,d^3x,$$

and S was an arbitrary closed surface, we conclude that

$$\nabla \cdot \mathbf{E} = 4\pi k\rho,$$

or, identifying $k = \frac{1}{4\pi\epsilon_0}$,

$$\nabla \cdot \mathbf{E} = \rho/\epsilon_0,$$

as advertised.

3.2.2 Absence of Magnetic Monopoles

Experiments have searched for magnetic monopoles, isolated North or South poles, but they have been inconclusive. Although there are theoretical grounds that might indicate that they do indeed exist,[1] we will not pursue this possibility further. In that case, the **B**-field is *solenoidal*, which is to say that it has zero divergence, and this is asserted by Maxwell's equation

$$\nabla \cdot \mathbf{B} = 0.$$

3.2.3 Ørsted and Ampère

That there was some sort of connection between electrical and magnetic phenomena is apparent from the erratic behavior of compasses in the vicinity of lightning strikes. However, the first experiments to provide unambiguous evidence for such a connection were performed by the Danish physicist Hans Christian Ørsted, following a chance observation (on 21 April 1819) while preparing a demonstration for his students to illustrate the heating of a wire carrying an electric current. He was also going to perform some demonstrations of magnetic phenomena for which he had available a compass, a magnetized needle balanced on a wooden stand. What Ørsted noticed was that the compass needle was deflected whenever the wire was connected to the battery, and returned to its original North-South orientation when the battery was disconnected. He followed this up with a series of experiments that confirmed that an electric current produces a magnetic field around the wire carrying the current. One may note that the study of phenomena associated with electric currents was made possible by the invention in 1800 by Count Alessandro Volta of the first electric battery.

Ørsted's results were reported to the Paris Academy of Sciences in 1820, and only a week after learning of this work André-Marie Ampère demonstrated further related research, in particular on the force between parallel wires carrying currents. The formal definition of the amp (or ampere, named after him), the unit of electric current, is based on this phenomenon. His *circuital law*, as formulated in 1826 may be expressed mathematically by

$$\oint_C \mathbf{H} \cdot d\mathbf{l} = I_{\text{enc}},$$

where I_{enc} is the net free current that passes through the closed loop C around which the line integral of the magnetic field **H** enters on the left-hand side of

---

[1]For example in P.A.M. Dirac "Quantised Singularities in the Electromagnetic Field," *Proc. Roy. Soc. Lond.* **A133**, 60 (1931), which used quantum mechanics to give an explanation for the quantization of electric charge. Grand unified theories and superstring theories also lend support.

the equation. This result holds only provided that the current is *steady* and the **H**-field is *static*.

3.2.4   The Law of Biot and Savart

Closely related to Ampère's circuital law is the *Biot-Savart* law, which dates from experiments performed by Jean-Baptiste Biot (1774–1862) and Felix Savart (1791–1841) in the 1820s. As with Ampère's law it applies to steady currents and static fields. It has the mathematical expression:

$$\mathbf{H} = \oint \frac{1}{4\pi} I \frac{d\mathbf{l} \times \mathbf{r}}{r^3}.$$

The integral is around a circuit carrying a current I.

The Biot-Savart law plays a role in *magnetostatics* similar to that played by Coulomb's law in electrostatics, and like it is analogous to Newton's law of universal gravitation. What Maxwell did was to take the crucial step needed to give a *dynamical* theory, no longer restricted to steady currents and static fields. That something new was needed becomes apparent if one applies Stokes's theorem to the Biot-Savart law so as to obtain $\int_S \nabla \times \mathbf{H} \cdot \mathbf{n} da = \int_S \mathbf{j} \cdot \mathbf{n} da$ for a surface S bounded by C with surface element $\mathbf{n} da$. This implies

$$\nabla \times \mathbf{H} = \mathbf{j},$$

which is, however, inconsistent with the local conservation law:

$$\frac{\partial \rho}{\partial t} + \nabla \cdot \mathbf{j} = 0.$$

What was required was an additional contribution to the current.

3.2.5   The Displacement Current

From the middle of the eighteenth century it was known that electric charge could be stored in what is now called a *capacitor*, in essence a pair of conductors ("plates") separated by a *dielectric* or insulating medium. As the capacitor is charged, a current flows into one of the plates, giving it a positive charge, and out from the other, leaving it with a negative charge. But no current flows between the plates, through the dielectric. Nevertheless, there is a displacement of charge within the dielectric, which changes as the charges on the plates change. It is this movement of charge that has to be accounted for in order to make Ampère's circuital law consistent with the conservation of charge. This was achieved by Maxwell in his 1861 paper "On Physical Lines of Force"[2] by adding to the "free" current $\mathbf{j}$ the *displacement current* $\frac{\partial \mathbf{D}}{\partial t}$, so as to give

$$\nabla \times \mathbf{H} = \mathbf{j} + \frac{\partial \mathbf{D}}{\partial t}.$$

---

[2]At Equation (112) in Part III of this publication (*Phil. Mag.* **21**, 338–348), in which Maxwell based his treatment on his theory of molecular vortices, which incorporated also the idea that electromagnetic phenomena are associated with a mechanical stress.

Maxwell's introduction of the displacement current was the crucial step needed to complete his grand edifice. Faraday had already postulated that there was energy stored in a charged capacitor, and that it resided in the dielectric between the plates. Maxwell took this idea forward and supposed that the dielectric was in some way stressed by the electric field, and this changing stress was described by the displacement current. It might be objected that there is no need for a dielectric medium to separate the plates of the capacitor; it could after all just be a vacuum between them. But for Maxwell, who believed in the existence of the *æther*, or "ether" as it is now more usually spelled, it was possible to extend his notion of charge separation in the dielectric to apply also to the ether. The ether theory has perished, but Maxwell's equations survive. And it is possible to find some lingering traces of the abandoned ether theory in the richness of properties ascribed to the vacuum in today's quantum field theory.[3]

To put any lingering doubts at rest, the magnetic field generated by a time-dependent displacement field **D** can be detected directly. In practical terms, this is most easily done with very rapidly changing fields, as exist in high-frequency applications, for example in FM radio with 10^9 Hz being typical.

3.2.6 Faraday's Law of Induction

Faraday's diary entry for August 29th, 1831 is introduced by

> 1. Expts. on the production of Electricty from Magnetism, etc.etc.

Over the next several months he conducted these experiments, from which he concluded that there was indeed an analog to the generation of a magnetic field by an electric current. But it required a *changing* magnetic flux through a circuit to produce an electrical effect. The change could be brought about by moving all or part of an electric circuit in the presence of a magnetic field, as in a dynamo. Or it could be brought about by changing the magnetic field with the circuit fixed, as for example in a transformer, where the changing magnetic field generated by changes in the current in the primary coil induces a current in the secondary coil. In either case the mathematical expression for the phenomenon, as given by Maxwell, again in "On Physical Lines of Force," is

$$\nabla \times \mathbf{E} = -\frac{\partial \mathbf{B}}{\partial t}.$$

3.2.7 The Lorentz Force

What is customarily described as the Lorentz force was in fact also given by Maxwell in "On Physical Lines of Force." The force on a charge q moving with velocity **v** in the presence of an electric field **E** and a magnetic field **B** is

$$\mathbf{F} = q(\mathbf{E} + \mathbf{v} \times \mathbf{B}). \tag{3.11}$$

---

[3]The vacuum is not empty, but filled with fields, which in quantum field theory cannot vanish because subject to fluctuations.

Together with the previous equations, it provides a basis for determining the electromotive force in either a circuit that changes shape or position in a fixed magnetic field, or a circuit that is fixed in a time-dependent magnetic field: or indeed in any situation when a charge or current experiences electric and magnetic fields. This equation provides the formal basis for the definition of the electric field and magnetic induction strengths, by relating them to the mechanical force on a charge. Its importance is also apparent in its widespread applications in situations as varied as the cathode ray tube (from the discovery of the electron to television), mass spectrometry, particle physics, and plasma physics, and indeed it pervades most of present-day physics.

3.3   Capacitors and Solenoids

The permittivity ϵ and permeability μ that enter into the constitutive equations, and their vacuum counterparts ϵ_0 and μ_0, have values that are related to experiments, of a kind performed by Faraday. Maxwell gave mathematical expression to these relations.

The simplest illustration comes from electrostatics, and from consideration of a parallel plate capacitor. Idealizing to a capacitor with large parallel plates, each of area A, distance d apart, with opposite charges $\pm Q$ on the plates, symmetry considerations allow one to argue that the electric field between the plates is uniform and perpendicular to the plates (one ignores the deviations that occur near the edges of the plates, since we have assumed that the plates are large). The electric potential difference between the plates is then $\Delta V = Ed$. The displacement field between the plates may be determined from the first of Maxwell's equations (Coulomb's law) to be $D = \epsilon E = \sigma$, the charge density on the plate, or $D = Q/A$. The capacitance is defined by $C = Q/\Delta V$, giving

$$C = \frac{\epsilon A}{d}, \qquad \epsilon = \frac{Q}{\Delta V}\frac{d}{A}.$$

Q and ΔV can both be measured (and likewise for A and d!). Their numerical values will of course depend on the *units* in terms of which they are expressed; the SI units are coulombs (C) for Q, volts (V) for V and farads (F) for the capacitance C.

The long cylindrical solenoidal coil plays a similar role to the parallel plate capacitor in consideration of the magnetic permeability. This time from Ampère's law one may conclude that the magnetic field H inside the coil when it is carrying a steady current I is $H = nI$, n being the number of turns per unit length of the solenoid. In terms of B, the corresponding magnetic induction, $B = \mu nI$, and B and I may be measured as can n, so determining $\mu = \frac{B}{nI}$. The unit for B is tesla (T) and for I ampere (amp) (1 A = 1 amp).

One has also to ask how coulomb, volt, amp, and tesla are defined. For the SI units we use, the starting point is the definition of the amp, which depends on Ampère's observation that there is a force between parallel current-carrying wires. The formal definition of the ampere is as follows:

The ampere is that constant current which, if maintained in two straight parallel conductors of infinite length, of negligible circular cross-section, and placed one meter apart in vacuum, would produce between these conductors a force equal to 2×10^{-7} newton per meter of length.[4]

The coulomb is then defined as the quantity of charge transported by a constant electric current of 1 amp flowing for 1 second (1 C = 1 A·s). The electric field (in volts per meter) and the magnetic field (in tesla) are then related to the force (again in newton) on a test charge through the Lorentz equation. The electric and magnetic units are thus established through relations with the mechanical units of the SI system.

The values μ_0 and ϵ_0 arise respectively for a solenoid with a vacuum inside the coil, or a capacitor with a vacuum between the plates. From the definition of the ampere, it follows that

$$\mu_0 = 4\pi \times 10^{-7} \frac{\text{N}}{\text{A}^2}. \tag{3.12}$$

As we will find below, the product $\epsilon_0 \mu_0$ equals c^{-2}, where c is the speed of light, a very well measured fundamental constant. In fact it is now used to *define* the meter in terms of the second by $c \equiv 2.99792458 \times 10^8$ m·s^{-1}, giving

$$\epsilon_0 \equiv \frac{1}{c^2 \mu_0} \approx 8.8542 \times 10^{-12} \frac{\text{F}}{\text{m}}. \tag{3.13}$$

The permittivity of a dielectric medium, ϵ, may be written as $\epsilon = \epsilon_r \epsilon_0$; ϵ_r, the *relative permittivity*, is often called the *dielectric constant* of the medium. Likewise the permeability of a medium, μ, may be written as $\mu = \mu_r \mu_0$, where the *relative permeability* μ_r is related to the *magnetic susceptibility* by $\chi_m = \mu_r - 1$.

3.3.1   Energy

From his experiments on capacitance and in partcular the dependence on the dielectric,[5] Faraday recognized that *energy* was stored in a charged capacitor, and related it to what he described as a stress in the space between the plates of the capacitor or the dielectric that occupied that space. Maxwell gave a mathematical representation for the *energy density* in that space in terms of the electric and displacement fields:

$$u_{\text{electric}} = \frac{1}{2} \mathbf{D} \cdot \mathbf{E} = \frac{1}{2} \epsilon \mathbf{E}^2. \tag{3.14}$$

For free space this becomes

$$u_{\text{electric}} = \frac{1}{2} \epsilon_0 \mathbf{E}^2.$$

---

[4]Bureau International des Poids et Mesures, SI Brochure (8th edition, 2006).
[5]The word was invented by Faraday.

For many dielectrics the dielectric constant ϵ_r is in the range $1 < \epsilon_r < 100$. Since the capacitance is increased when a dielectric is inserted in the space between the plates of a capacitor, it requires a force to pull the dielectric out. Put another way, a dielectric is attracted to regions of high electric field, and this is exploited in what are called *optical tweezers* that use a focused laser beam to manipulate micron-sized dielectric particles.

Faraday also studied the energy stored in a magnetic field; for example that produced in a solenoidal coil carrying an electric current. The field, and so also the energy, is increased if the core of the solenoid contains a suitable medium, for example iron. Indeed an iron rod will be attracted into the core of the solenoid and requires a force to extract it. Maxwell put that too into a mathematical expression, this time for the magnetic energy density:

$$u_{\text{magnetic}} = \frac{1}{2}\mathbf{H} \cdot \mathbf{B} = \frac{1}{2}\frac{1}{\mu}\mathbf{B}^2. \tag{3.15}$$

For free space this becomes

$$u_{\text{magnetic}} = \frac{1}{2}\frac{1}{\mu_0}\mathbf{B}^2.$$

3.4 Electromagnetic Waves

In the vacuum, in the absence of sources, Maxwell's equations are

$$
\begin{aligned}
\nabla \cdot \mathbf{E} &= 0, \\
\nabla \times \mathbf{B} &= \epsilon_0\mu_0\frac{\partial \mathbf{E}}{\partial t}, \\
\nabla \cdot \mathbf{B} &= 0, \\
\nabla \times \mathbf{E} &= -\frac{\partial \mathbf{B}}{\partial t}.
\end{aligned}
\tag{3.16}
$$

From these it follows that

$$
\begin{aligned}
\nabla^2\mathbf{E} - \frac{1}{c^2}\frac{\partial^2 \mathbf{E}}{\partial t^2} &= 0, \\
\nabla^2\mathbf{B} - \frac{1}{c^2}\frac{\partial^2 \mathbf{B}}{\partial t^2} &= 0,
\end{aligned}
\tag{3.17}
$$

where

$$c^2 = \frac{1}{\epsilon_0\mu_0}. \tag{3.18}$$

That is, each component of the fields $\mathbf{E}$ and $\mathbf{B}$ satisfies the *wave equation*.

3.4.1 Polarization

Electromagnetic waves are *transverse*. What this means is that both the electric field vector $\mathbf{E}$ and the magnetic induction vector $\mathbf{B}$ are perpendicular to the *wave-vector* $\mathbf{k}$, which points along the direction of propagation of the wave.[6] The general solution to the wave equation for $\mathbf{E}$ is a linear superposition of *plane waves* of form

$$\mathbf{E} = \mathbf{E}_0 \cos(\mathbf{k} \cdot \mathbf{x} - \omega t)$$

with angular frequency $\omega = ck$ and wavelength $\lambda = 2\pi/k$ traveling in the direction of $\mathbf{k}$ ($k = |\mathbf{k}|$). The first of Maxwell's equations, $\nabla \cdot \mathbf{E} = 0$ then gives $\mathbf{k} \cdot \mathbf{E}_0 = 0$. If we write $\mathbf{E}_0 = E_0\mathbf{n}$, there are *two* independent unit vectors $\mathbf{n}$ that satisfy the orthogonality condition, and they are used to characterize the *polarization* of the wave. From

$$\dot{\mathbf{B}} = -\nabla \times \mathbf{E} = -\mathbf{E}_0 \times \mathbf{k} \sin(\mathbf{k} \cdot \mathbf{x} - \omega t)$$

it follows that

$$\mathbf{B} = \mathbf{B}_0 \cos(\mathbf{k} \cdot \mathbf{x} - \omega t)$$

with $\mathbf{B}_0 = \frac{1}{ck}\mathbf{E}_0 \times \mathbf{k}$. The vectors $\mathbf{k}, \mathbf{E}, \mathbf{B}$ are mutually orthogonal.

3.4.2 Electromagnetism and Light

The speed c of the propagation of electromagnetic waves is, of course, the speed of light. We may now say "of course", but it was Maxwell's recognition that his equations entail the prediction of electromagnetic waves, and that their speed can be calculated from the constants ϵ_0 and μ_0 related to electric and magnetic phenomena that allowed him to write in his paper "A Dynamical Theory of the Electromagnetic Field"[7]

> The agreement of the results [his calculation of the speed of propogation of electromagnetic waves with what had been measured as the speed of light] seems to show that light and magnetism are affections of the same substance, and that light is an electromagnetic disturbance propagated through the field according to electromagnetic laws[8].

The independence of the speed of propagation of light from the motion of the source (and also of the detector) is inconsistent with Galilean relativity, and was a source of much debate in the years at the end of the nineteenth century. It was the stimulus for Einstein's revolutionary theory of relativity,

---

[6]This applies only to the space- and time-dependent contributions to the fields; constant static electric and magnetic fields can also be present, and are not restricted by these cosiderations which apply only to the propagating components of the fields.

[7]*Phil. Trans. Roy. Soc.* **155**, 459–512 (1865).

[8]Maxwell later himself devised and perormed an experiment that gave a direct measurement of what we would now write as $1/\sqrt{\epsilon_0\mu_0}$. See "On a Method of Making a Direct Comparison of Electrostatic with Electromagnetic Force; with a Note on the Electromagnetic Theory of Light," *Phil. Trans. Roy. Soc.* **158**, 643–658 (1868).

published as one of his *annus mirabilis* papers of 1905, "On the Electrodynamics of Moving Bodies."[9] It is impossible to improve on the clarity with which Einstein introduces his ideas, so that I am obliged to quote in full the opening paragraphs of that paper:

> It is known that Maxwell's electrodynamics—as usually understood at the present time—when applied to moving bodies, leads to asymmetries which do not appear to be inherent in the phenomena. Take, for example, the reciprocal electrodynamic action of a magnet and a conductor. The observable phenomenon here depends only on the relative motion of the conductor and the magnet, whereas the customary view draws a sharp distinction between the two cases in which either the one or the other of these bodies is in motion. For if the magnet is in motion and the conductor at rest, there arises in the neighbourhood of the magnet an electric field with a certain definite energy, producing a current at the places where parts of the conductor are situated. But if the magnet is stationary and the conductor in motion, no electric field arises in the neighbourhood of the magnet. In the conductor, however, we find an electromotive force, to which in itself there is no corresponding energy, but which gives rise—assuming equality of relative motion in the two cases discussed—to electric currents of the same path and intensity as those produced by the electric forces in the former case.
>
> Examples of this sort, together with the unsuccessful attempts to discover any motion of the earth relatively to the "light medium," suggest that the phenomena of electrodynamics as well as of mechanics possess no properties corresponding to the idea of absolute rest. They suggest rather that, as has already been shown to the first order of small quantities, the same laws of electrodynamics and optics will be valid for all frames of reference for which the equations of mechanics hold good. We will raise this conjecture (the purport of which will hereafter be called the "Principle of Relativity") to the status of a postulate, and also introduce another postulate, which is only apparently irreconcilable with the former, namely, that light is always propagated in empty space with a definite velocity c which is independent of the state of motion of the emitting body. These two postulates suffice for the attainment of a simple and consistent theory of the electrodynamics of moving bodies based on Maxwell's theory for stationary bodies. The introduction of a "luminiferous ether" will prove to be superfluous inasmuch as the view here to be developed will not require an "absolutely stationary space" provided with special properties, nor assign a velocity-vector to a point of the empty space in which electromagnetic processes take place.

---

[9] *Ann. Phys.* **XVII** (1905): reprinted in English translation in his book *The Principle of Relativity* (Dover Publications, New York, 1952).

We will recast Maxwell's equations in a formulation that brings to the fore their consistency with Einstein's principle of relativity.

3.5   Exercises for Chapter 3

1   Consider a parallel plate capacitor with plates of area A and separation x, with a vacuum between the plates. If the charge on one plate is Q and on the other $-Q$, and the dimensions are such that edge effects may be neglected, what is the magnitude of electric field E in the region between the plates? What is its direction? What is the force exerted by the one plate on the other? How much work must be done in order to increase the separation by an amount dx? How much energy is stored in the capacitor when the plates are separated by x? If this energy resides in the electric field, express this energy in terms of the volume of the region in which the electric field exists, and the magnitude of the field there. Hence show that the energy density associated with an electric field in the vacuum is

$$u_E = \frac{1}{2}\epsilon_0 E^2.$$

2   Consider now a long straight solenoidal coil, with n turns of wire per unit length, carrying a current I. The cross-sectional area of the solenoid is S. What is the inductance of a length l of the solenoid? Neglecting end effects, what is the magnitude of the magnetic induction B in the region inside the solenoid? If the solenoid, which has a total inductance L, is connected in series with a source $\mathcal{E}$ of e.m.f. and a resistor, the total ohmic rsistance of the circuit being R, show that

$$\mathcal{E}I = I^2 R + LI\frac{dI}{dt},$$

and hence show that if U_B is the energy stored in the magnetic field in the solenoid, $dU_B = LI\,dI$. Use this to derive $U_B = \frac{1}{2}LI^2$ and hence obtain

$$u_B = \frac{1}{2}\frac{B^2}{\mu_0}$$

for the energy density in the magnetic field.

3   Explain how the energy density of the elctromagnetic field, namely,

$$u = u_E + u_B = \frac{1}{2}(\epsilon_0 E^2 + B^2/\mu_0),$$

is modified in the presence of media, so as to give

$$u = \frac{1}{2}(\mathbf{E}\cdot\mathbf{D} + \mathbf{B}\cdot\mathbf{H}).$$

Chapter 4

Behavior under Lorentz Transformations

In this chapter we will show how the electric and magnetic fields behave under Lorentz transformations. This will confirm that the electric and magnetic fields are intimately related, and that they are best regarded as parts of one physical entity, the electromagnetic field. To do this it is expedient to use the language of four-vectors and tensors introduced in Chapter 2. The behavior of the fields under rotations is already implicit in the use of the vector notation in the previous chapter, but we need now to consider boosts, the Lorentz transformations that relate properties in an inertial frame K to those in a frame K' moving with constant velocity with respect to K in the "standard orientation" as described in Chapter 2.

4.1   The Charge-Current Density Four-Vector

We will start by discussing the *sources* of the fields, namely, the charge and current densities, ρ and $\mathbf{j}$. Consider an element of charge at rest in the frame K, occupying the volume element $dx^1 \, dx^2 \, dx^3$, so that the charge contained in that volume is

$$dq = \rho \, dx^1 \, dx^2 \, dx^3.$$

In the frame K' this same element of charge will occupy the volume element $dx'^1 \, dx'^2 \, dx'^3 = (\gamma^{-1} dx^1) \, dx^2 \, dx^3$ and since the total amount of charge in the volume element is the same in both frames,

$$dq = \rho' \, dx'^1 \, dx'^2 \, dx'^3,$$

and it follows that

$$\rho' = \gamma \rho. \tag{4.1}$$

But note also that in K' the element of charge is *moving*, with a velocity $\mathbf{u}' = -\beta c$, so that there is also a current density in K' given by

$$\mathbf{j}' = \rho' \mathbf{u}' = -\gamma \beta c \rho. \tag{4.2}$$

So we have $(\rho; \mathbf{j} = 0)$ in K transforming to $(\rho' = \gamma\rho; \mathbf{j}' = -\gamma\beta c\rho)$ in K', which suggests writing $j^0 = c\rho, \mathbf{j} = (j^1, j^2, j^3)$ since then the transformation of j^μ may be recognized as that of a four-vector. In fact our element of charge has a four-velocity U^μ, and we have

$$j^\mu = \rho_0 U^\mu, \tag{4.3}$$

where ρ_0 is the charge density in the frame in which the element of charge is at rest. We may conclude that the charge-current density $j = (c\rho, \mathbf{j})$ is a contravariant four-vector.

Note also that the continuity equation

$$\frac{\partial \rho}{\partial t} + \nabla \cdot \mathbf{j} = 0$$

is

$$\frac{\partial j^0}{\partial x^0} + \frac{\partial j^1}{\partial x^1} + \frac{\partial j^2}{\partial x^2} + \frac{\partial j^3}{\partial x^3} = 0$$

or

$$\partial_\mu j^\mu = 0, \tag{4.4}$$

where

$$\partial_\mu \equiv \frac{\partial}{\partial x^\mu} \quad \left(\partial_0 = \frac{\partial}{c\partial t}; \quad \frac{\partial}{\partial_k}, \ k = 1, 2, 3 \right).$$

The continuity equation is thus a manifestly covariant equation.

4.2   The Lorentz Force

Having found the transformation rule for the *sources* j^μ, it is natural to turn to the question of the transformation rules for the fields themselves. As a step in this direction, recall that the fields are in fact related through the Lorentz force equation to the mechanical force as determined by the change in momentum of a test particle of charge q moving with velocity $\mathbf{u}$ and momentum $\mathbf{p}$:

$$d\mathbf{p}/dt = \mathbf{F} = q(\mathbf{E} + \mathbf{u} \times \mathbf{B}).$$

Thus we have

$$\begin{aligned}
\frac{d\mathbf{p}}{d\tau} &= q(\mathbf{E} + \mathbf{u} \times \mathbf{B})\frac{dt}{d\tau} \\
&= q(\mathbf{E} + \mathbf{u} \times \mathbf{B})\gamma_u \\
&= q(\frac{1}{c}\mathbf{E}U^0 + \mathbf{U} \times \mathbf{B}),
\end{aligned} \tag{4.5}$$

and also the rate of change of the energy $p^0 c$ of the test particle follows from

$$\begin{aligned}
\frac{dp^0}{d\tau} &= \frac{1}{c}\frac{dt}{d\tau}\frac{dp^0 c}{dt} \\
&= \frac{1}{c}\gamma q \mathbf{E} \cdot \mathbf{u} \\
&= \frac{1}{c}\mathbf{E} \cdot \mathbf{U}.
\end{aligned} \tag{4.6}$$

So

$$\frac{d}{d\tau}\begin{pmatrix} p^0 \\ \mathbf{p} \end{pmatrix} = \frac{q}{c}\begin{pmatrix} \mathbf{E} \cdot \mathbf{U} \\ EU^0 + c\mathbf{U} \times \mathbf{B} \end{pmatrix}$$

which may be written as

$$\frac{d}{d\tau}p^\alpha = qF^{\alpha\beta}U_\beta, \tag{4.7}$$

where $F^{\alpha\beta}$ is a certain second-rank tensor, the components of which are given in terms of the $\mathbf{E}$ and $\mathbf{B}$ fields. It is possible to obtain this result directly, but it is easier and also more useful to proceed by introducing the *potential four-vector*.

4.3   The Potential Four-Vector

Already in his "A Dynamical Theory of the Electromagnetic Field" Maxwell had introduced what he called the electromagnetic momentum[1] and the electric potential, which in our notation are respectively $\mathbf{A}$ and Φ, the vector potential and the scalar electric potential.

Using (2.13), the third equation of (3.9) is satisfied if we write

$$\mathbf{B} = \boldsymbol{\nabla} \times \mathbf{A}. \tag{4.8}$$

The fourth equation now becomes $\boldsymbol{\nabla} \cdot \left(\mathbf{E} + \frac{\partial \mathbf{A}}{\partial t}\right)$, and from (2.14) this is satisfied by

$$\mathbf{E} = -\boldsymbol{\nabla}\Phi - \frac{\partial \mathbf{A}}{\partial t}. \tag{4.9}$$

The remaining two equations from the set of equations (3.9), which involve the sources $\mathbf{j}$ and ρ, become

$$\boldsymbol{\nabla} \cdot \mathbf{E} = \boldsymbol{\nabla} \cdot (-\boldsymbol{\nabla}\Phi - \dot{\mathbf{A}})$$
$$= -\nabla^2\Phi - \boldsymbol{\nabla} \cdot \dot{\mathbf{A}}$$
$$= \Box\Phi - \frac{\partial}{\partial t}\left(\frac{1}{c^2}\dot{\Phi} + \boldsymbol{\nabla} \cdot \mathbf{A}\right)$$
$$= \rho/\epsilon_0, \tag{4.10}$$

and with the use of the vector identity

$$\boldsymbol{\nabla} \times (\boldsymbol{\nabla} \times \mathbf{A}) = \boldsymbol{\nabla}(\boldsymbol{\nabla} \cdot \mathbf{A}) - \nabla^2\mathbf{A},$$

---

[1] Maxwell's choice of the name "electromagnetic potential" was motivated by his mechanical model for the electromagnetic field, and since force is rate of change of momentum, it is not surprising that its time derivative enters in his expression for the electromotive force. It was also prescient insofar as the canonical momentum for a charged particle in the presence of an electromagnetic field receives a contribution from this quantity as well as the mechanical momentum $\mathbf{p}$. Faraday too had hypothesized what he called the "electro-tonic state," relating it to the stress he postulated as a property of the space surrounding electric currents or magnets.

$$\nabla \times \mathbf{B} - \epsilon_0 \mu_0 \frac{\partial \mathbf{E}}{\partial t} = \nabla \times (\nabla \times \mathbf{A}) + \frac{1}{c^2}\left(\nabla \dot{\Phi} + \ddot{\mathbf{A}}\right)$$

$$= \Box \mathbf{A} + \nabla \left(\frac{1}{c^2}\dot{\Phi} + \nabla \cdot \mathbf{A}\right)$$

$$= \mu_0 \mathbf{j}. \tag{4.11}$$

Here again the symbol $\Box$ is used for the Lorentz-invariant d'Alembertian operator

$$\Box \equiv \frac{1}{c^2}\frac{\partial^2}{(\partial t)^2} - \nabla^2 = \partial^\mu \partial_\mu,$$

where $\partial^\mu = \eta^{\mu\nu}\partial_\nu$.

We have already seen that the sources may be put together as a contravariant four-vector $j = (c\rho, \mathbf{j})$, so we may write

$$j^0 = c\epsilon_0 \left(\Box \Phi - \frac{\partial}{\partial t}\left(\frac{1}{c^2}\dot{\Phi} + \nabla \cdot \mathbf{A}\right)\right),$$

$$\mathbf{j} = \frac{1}{\mu_0}\left(\Box \mathbf{A} + \nabla \left(\frac{1}{c^2}\dot{\Phi} + \nabla \cdot \mathbf{A}\right)\right).$$

These may be written more succinctly as

$$j^\mu = \frac{1}{\mu_0}(\Box A^\mu - \partial^\mu \Lambda) \tag{4.12}$$

where we have used $c^2 \epsilon_0 \mu_0 = 1$, and define $A^0 = \Phi/c$. Also

$$\Lambda \equiv \left(\frac{1}{c^2}\dot{\Phi} + \nabla \cdot \mathbf{A}\right) = \partial_\mu A^\mu. \tag{4.13}$$

Since we know that the components of j^μ transform as a four-vector, consistency requires that the same is true for A^μ, which also makes Λ a Lorentz scalar.

4.4   Gauge Transformations

In developing his notion of the electro-tonic state, Faraday was aware that it was only the *changes* in that quantity that were of physical significance, not its absolute magnitude. Maxwell too, who called it the "electromagnetic momentum," also remarked on this. He also stated that it is a vector quantity and used for it the letter " A" from which derives our notation $\mathbf{A}$, and it is in fact the vector potential. And he recognized that from what we would write as

$$\nabla \cdot \mathbf{B} = 0,$$

it follows that there exists $\mathbf{A}$ such that

$$\mathbf{B} = \nabla \times \mathbf{A}.$$

Of course $\mathbf{A}$ is not uniquely determined, since we can change it by

$$\mathbf{A} \to \mathbf{A}' = \mathbf{A} + \nabla \chi$$

with χ a scalar function, and this will leave $\mathbf{B}$ unchanged, because (see (2.14))

$$\boldsymbol{\nabla} \times (\mathbf{A}' - \mathbf{A}) = 0.$$

But if $\mathbf{A}$ has been replaced by $\mathbf{A} + \boldsymbol{\nabla}\chi$ where χ is an arbitrary function of space and time, what is the consequence for the law of induction,

$$\boldsymbol{\nabla} \times \mathbf{E} + \frac{\partial \mathbf{B}}{\partial t} = 0,$$

which we had said in (4.9) followed from

$$\mathbf{E} + \frac{\partial \mathbf{A}}{\partial t} = -\boldsymbol{\nabla}\Phi,$$

for some function Φ? If we now change $\mathbf{A}$ according to

$$\mathbf{A} \rightarrow \mathbf{A} + \boldsymbol{\nabla}\chi \tag{4.14}$$

we must take

$$\Phi \rightarrow \Phi - \frac{\partial}{\partial t}\chi \tag{4.15}$$

in order that $\mathbf{E}$ should be left unchanged. Thus for any *potentials* $\mathbf{A}, \Phi$ with the definitions

$$\begin{aligned} \mathbf{B} &= \boldsymbol{\nabla} \times \mathbf{A}, \\ \mathbf{E} &= -\boldsymbol{\nabla}\Phi - \frac{\partial \mathbf{A}}{\partial t}, \end{aligned} \tag{4.16}$$

two of Maxwell's equations, namely, the pair

$$\begin{aligned} \boldsymbol{\nabla} \cdot \mathbf{B} &= 0, \\ \boldsymbol{\nabla} \times \mathbf{E} + \frac{\partial \mathbf{B}}{\partial t} &= 0 \end{aligned} \tag{4.17}$$

are automatically satisfied. Furthermore, the simultaneous change

$$\begin{aligned} \mathbf{A} &\rightarrow \mathbf{A} + \boldsymbol{\nabla}\chi, \\ \Phi &\rightarrow \Phi - \frac{\partial \chi}{\partial t} \end{aligned} \tag{4.18}$$

to the vector and scalar potentials, which is called a *gauge transformation*,[2] leaves the fields $\mathbf{E}$ and $\mathbf{B}$ unaltered. They are *gauge invariant*.

Under a gauge transformation, the scalar quantity Λ previously defined transforms as

$$\Lambda \rightarrow \Lambda' = \Lambda - \Box\chi.$$

---

[2] The name comes from Hermann Weyl's 1918 book *Raum, Zeit, Materie* in which he attempted to unify Einstein's general relativity with Maxwell's electromagnetism, using what he called a change in the *gauge* or scale to extend the general coordinate invariance of general relativity. Although this theory was flawed, the change in the electromagnetic potentials that he used was precisely the gauge transformation we are now discussing.

By suitable choice of χ it is thus apparent that it is always possible to impose the condition that

$$\Lambda \equiv \frac{1}{c^2}\frac{\partial \Phi}{\partial t} + \boldsymbol{\nabla} \cdot \mathbf{A} = \partial_\mu A^\mu = 0, \tag{4.19}$$

which is known as the *Lorenz gauge*[3] condition. This condition is indeed Lorentz invariant: if true in one frame, it remains satisfied in any other.

An alternative possible choice of gauge, useful for some problems, is one in which $\boldsymbol{\nabla} \cdot \mathbf{A} = 0$; this is called a *Coulomb, radiation,* or *transverse* gauge. (This gauge choice is not in general Lorentz invariant.) Other gauge conditions are sometimes imposed.

Whatever the choice of gauge, as we have seen, the pair of Maxwell's equations which do not involve the source terms are automatically satisfied. We now turn to the other pair of equations, namely,

$$\boldsymbol{\nabla} \cdot \mathbf{E} = \mu_0 c^2 \rho,$$

$$\boldsymbol{\nabla} \times \mathbf{B} = \mu_0 \left(\mathbf{j} + \epsilon_0 \frac{\partial \mathbf{E}}{\partial t} \right).$$

Written in terms of the potentials, and again making use of the vector identity

$$\boldsymbol{\nabla} \times (\boldsymbol{\nabla} \times \mathbf{A}) = \boldsymbol{\nabla}(\boldsymbol{\nabla} \cdot \mathbf{A}) - \nabla^2 \mathbf{A},$$

these become

$$\Box \mathbf{A} + \boldsymbol{\nabla} \left(\boldsymbol{\nabla} \cdot \mathbf{A} + \frac{1}{c^2}\dot{\Phi} \right) = \mu_0 \mathbf{j},$$

$$\Box \frac{1}{c}\Phi - \frac{1}{c}\frac{\partial}{\partial t} \left(\boldsymbol{\nabla} \cdot \mathbf{A} + \frac{1}{c^2}\dot{\Phi} \right) = \mu_0 c \rho. \tag{4.20}$$

Recalling that we have defined

$$\Lambda \equiv \boldsymbol{\nabla} \cdot \mathbf{A} + \frac{1}{c^2}\dot{\Phi},$$

and supposing that we define

$$A^0 \equiv \frac{1}{c}\Phi; \quad \mathbf{A} = (A_x, A_y, A_z) = (A^1, A^2, A^3), \tag{4.21}$$

what we have is

$$\Box A^\mu - \eta^{\mu\nu}\partial_\nu \Lambda = \mu_0 j^\mu.$$

And since we know that j^μ is a four-vector, this confirms that A^μ is likewise a four-vector, in which case Λ is indeed a Lorentz scalar, and the field equations satisfied by the potentials A^μ are seen to be Lorentz covariant.

Note that if we impose the Lorenz gauge condition, so putting $\Lambda = 0$, the equations simplify to

$$\Box A^\mu = \mu_0 j^\mu, \qquad \partial_\mu A^\mu = 0. \tag{4.22}$$

---

[3]Named after Ludvig Lorenz. Often mis-spelled Lorentz gauge, from confusion with Hendrik Lorentz.

4.5 The Field-Strength Tensor

We have expressed the electric and magnetic fields in terms of the potentials, and we know how the potentials change under a Lorentz transformation,

$$A^\mu \to A'^\mu = \Lambda^\mu{}_\nu A^\nu,$$

where $\Lambda^\mu{}_\nu$ is the transformation matrix introduced in Section 2.4. From this follows the transformation laws for the electric and magnetic fields. To obtain them most simply, note that the field components are all of them of the form

$$F^{\alpha\beta} \equiv \partial^\alpha A^\beta - \partial^\beta A^\alpha,$$

where

$$\partial^\alpha \equiv \eta^{\alpha\beta}\partial_\beta \Rightarrow \partial^0 = \partial_0; \quad \partial^1 = -\partial_1, \ etc.$$

For example,

$$B_x = \frac{\partial A_z}{\partial y} - \frac{\partial A_y}{\partial z} = \partial^2 A^3 - \partial^3 A^2 = F^{23}$$

and

$$\frac{1}{c}E_x = -\frac{1}{c}\frac{\partial \Phi}{\partial x} - \frac{1}{c}\frac{\partial A_x}{\partial t} = -(\partial^0 A^1 - \partial^1 A^0) = -F^{01}.$$

Thus the components of the tensor $F^{\alpha\beta}$ arranged as a matrix are

$$F^{\alpha\beta} = \begin{pmatrix} 0 & -E_x/c & -E_y/c & -E_z/c \\ E_x/c & 0 & B_z & -B_y \\ E_y/c & -B_z & 0 & B_x \\ E_z/c & B_y & -B_x & 0 \end{pmatrix}. \tag{4.23}$$

This second-rank contravariant antisymmetrical tensor is the *field-strength tensor*; because of its antisymmetry $F^{\alpha\beta} = -F^{\beta\alpha}$, it has only six independent components. We can of course lower its indices to obtain $F_{\alpha\beta} = \eta_{\alpha\mu}\eta_{\beta\nu}F^{\mu\nu}$, with

$$F_{\alpha\beta} = \begin{pmatrix} 0 & E_x/c & E_y/c & E_z/c \\ -E_x/c & 0 & B_z & -B_y \\ -E_y/c & -B_z & 0 & B_x \\ -E_z/c & B_y & -B_x & 0 \end{pmatrix}. \tag{4.24}$$

For any choice of potentials A_μ, the fields $F_{\mu\nu} = \partial_\mu A_\nu - \partial_\nu A_\mu$ automatically satisfy

$$\partial_\lambda F_{\mu\nu} + \partial_\mu F_{\nu\lambda} + \partial_\nu F_{\lambda\mu} = 0, \tag{4.25}$$

and expressed in terms of the electric and magnetic fields, these equations are once again the homogeneous pair of Maxwell's equations, that is, those that do not involve the sources.

As for the inhomogeneous pair of Maxwell's equations, these are obtained from

$$\begin{aligned} \partial_\alpha F^{\alpha\beta} &= \partial_\alpha(\partial^\alpha A^\beta - \partial^\beta A^\alpha) \\ &= \Box A^\beta - \partial^\beta(\partial_\alpha A^\alpha) \\ &= \Box A^\beta - \partial^\beta \Lambda \\ &= \mu_0 j^\beta. \end{aligned} \tag{4.26}$$

The field-strength tensor is of course gauge invariant; its components are after all expressed in terms of the fields $\mathbf{E}$ and $\mathbf{B}$.

Returning for a moment to the force equation, for which we had

$$\frac{d}{d\tau}\begin{pmatrix} p^0 \\ \mathbf{p} \end{pmatrix} = \frac{q}{c}\begin{pmatrix} \mathbf{E}\cdot\mathbf{U} \\ \mathbf{E}U^0 + c\mathbf{U}\times\mathbf{B} \end{pmatrix},$$

we now find that

$$\begin{aligned}
\frac{d}{d\tau}p^0 &= \frac{q}{c}(E_x U_x + E_y U_y + E_z U_z) \\
&= q(-F^{01}U^1 - F^{02}U^2 - F^{03}U^3) \\
&= q(F^{01}U_1 + F^{02}U_2 + F^{03}U_3) \\
&= qF^{0\beta}U_\beta
\end{aligned} \tag{4.27}$$

and for example

$$\begin{aligned}
\frac{d}{d\tau}p^1 &= q(-F^{01}U^0 - U^2 F^{12} + U^3 F^{13}) \\
&= q(F^{10}U_0 + F^{12}U_2 + F^{13}U_3) \\
&= qF^{1\beta}U_\beta.
\end{aligned} \tag{4.28}$$

So, as advertised,

$$\frac{d}{d\tau}p^\alpha = qF^{\alpha\beta}U_\beta. \tag{4.29}$$

To spell out what precisely *is* the Lorentz transformation law for the electromagnetic field, we recall that the transformation law for any contravariant tensor such as $F^{\mu\nu}$ is given by

$$F^{\mu\nu} \to F'^{\mu\nu} = \Lambda^\mu{}_\kappa \Lambda^\nu{}_\lambda F^{\kappa\lambda}$$

where for the standard boost the coefficients $\Lambda^\mu{}_\kappa$ may be arranged as the matrix

$$\begin{pmatrix} \cosh\zeta & -\sinh\zeta & 0 & 0 \\ -\sinh\zeta & \cosh\zeta & 0 & 0 \\ 0 & 0 & 1 & 0 \\ 0 & 0 & 0 & 1 \end{pmatrix}.$$

So in terms of matrices, we have

$$F \to F' = \Lambda F \Lambda^T \tag{4.30}$$

(Λ^T denoting the matrix transpose of Λ). Straightforward matrix multiplication then yields

$$E'_x = E_x,$$
$$E'_y = E_y \cosh\zeta - B_z \, c \sinh\zeta,$$
$$E'_z = E_z \cosh\zeta + B_y \, c \sinh\zeta; \tag{4.31}$$

$$B'_x = B_x,$$
$$B'_y = B_y \cosh\zeta + \frac{E_z}{c}\sinh\zeta,$$
$$B'_z = B_z \cosh\zeta - \frac{E_y}{c}\sinh\zeta. \tag{4.32}$$

These can also be written as

$$\mathbf{E}' = \gamma(\mathbf{E} + c\boldsymbol{\beta}\times\mathbf{B}) - \frac{\gamma^2}{\gamma+1}\boldsymbol{\beta}(\boldsymbol{\beta}\cdot\mathbf{E}),$$
$$\mathbf{B}' = \gamma(\mathbf{B} - \boldsymbol{\beta}\times\mathbf{E}/c) - \frac{\gamma^2}{\gamma+1}\boldsymbol{\beta}(\boldsymbol{\beta}\cdot\mathbf{B}). \tag{4.33}$$

4.6   The Dual Field-Strength Tensor

Another useful tensor whose components are the electric and magnetic field strengths is the *dual* tensor defined by

$$^*F_{\alpha\beta} = \frac{1}{2}\epsilon_{\alpha\beta\mu\nu}F^{\mu\nu} \tag{4.34}$$

which makes use of the *Levi-Civita alternating tensor* $\epsilon_{\alpha\beta\mu\nu}$:

$$\epsilon_{\alpha\beta\mu\nu} = \begin{cases} \pm 1, & \text{if } \alpha\beta\mu\nu \text{ is a } \pm\text{ve permutation of 0123;} \\ 0, & \text{otherwise.} \end{cases} \tag{4.35}$$

Thus

$$^*F_{01} = \frac{1}{2}(\epsilon_{0123}F^{23} + \epsilon_{0132}F^{32}) = F^{23} = -B_x, \text{etc.}$$

and

$$^*F_{12} = \frac{1}{2}(\epsilon_{1203}F^{03} + \epsilon_{1230}F^{30}) = F^{03} = -E_z/c, \text{etc.,}$$

so that

$$^*F_{\alpha\beta} = \begin{pmatrix} 0 & -B_x & -B_y & -B_z \\ B_x & 0 & -E_z/c & E_y/c \\ B_y & E_z/c & 0 & -E_x/c \\ B_z & -E_y/c & E_x/c & 0 \end{pmatrix}. \tag{4.36}$$

To show that $^*F$ is in fact a tensor, it is simplest to show that $\epsilon_{\alpha\beta\mu\nu}$ is a tensor, and this is done by consideration of the formula

$$\epsilon_{\alpha\beta\mu\nu} \to \epsilon'_{\alpha\beta\mu\nu} = \frac{\partial x^\gamma}{\partial x'^\alpha} \frac{\partial x^\delta}{\partial x'^\beta} \frac{\partial x^\rho}{\partial x'^\mu} \frac{\partial x^\sigma}{\partial x'^\nu} \epsilon_{\gamma\delta\rho\sigma}$$

$$= \det\left(\frac{\partial x}{\partial x'}\right) \epsilon_{\alpha\beta\mu\nu}. \tag{4.37}$$

The determinant is just $[\det \Lambda]^{-1}$, which from the result obtained earlier is ± 1. Its presence shows that ϵ is in fact a *pseudotensor*; it changes sign under a reflection or a change in the sense of time. Lorentz transformations like these, which have $\det \Lambda = -1$ are called *improper*, and we will henceforth exclude them. For the *proper* Lorentz transformations, the determinant factor is unity.

It is interesting to note that the relationship between F and $^*F$ is one that interchanges $\mathbf{B} \longleftrightarrow -\mathbf{E}$.[4]

Since F and $^*F$ are both tensors, the contractions $F^{\mu\nu}F_{\mu\nu}, F^{\mu\nu}{}^*F_{\mu\nu}$ and $^*F^{\mu\nu}{}^*F_{\mu\nu}$ are Lorentz *scalars*. It is easy to evaluate them; they are repectively $2(\mathbf{B}^2 - \mathbf{E}^2/c^2), 4\mathbf{E} \cdot \mathbf{B}/c$, and $2(\mathbf{E}^2/c^2 - \mathbf{B}^2)$. In fact these are the *only* independent Lorentz scalars that can be constructed from the field strengths. Note that they allow a classification of electromagnetic field configurations according as $E^2 > B^2c^2$, $E^2 = B^2c^2$, $E^2 < B^2c^2$ which is invariant under Lorentz transformations.

It is also evident that the tensors F and $^*F$ are gauge invariant; their components are after all given in terms of the fields $\mathbf{E}$ and $\mathbf{B}$.

It turns out, as we will show in what follows, that the contraction $F^{\mu\nu}F_{\mu\nu}$ will provide the Lagrangian for the electromagnetic field. But first let us review the Lagrange method for the dynamics of particles.

4.7   Exercises for Chapter 4

1   (a)   Write down the equation for the Lorentz force on a particle of charge q moving with velocity $\mathbf{v}$. Explain how this may be generalized to the equation

$$\mathbf{f} = \rho\mathbf{E} + \mathbf{j} \times \mathbf{B}$$

for the force per unit volume acting on a charge and current density.
   (b)   Show that

$$f^k = F^{k\alpha}j_\alpha \qquad \text{for} \quad k = 1, 2, 3,$$

and define f^0 so that f^α is a four-vector.
   (c)   Give a physical interpretation for f^α.

---

[4]This is an illustration of a *duality* between the electric and magnetic fields, that perhaps lends support to the conjecture that magnetic monopoles might exist. Duality relations have an important role in what is known as *M-theory*, a theory that extends *string theory*, which is anticipated to provide a consistent unified quantum theory of gravity and the *standard model* of particle physics.

2 (a) Show that potentials of the form

$$A^\mu = a^\mu \, \cos[\omega t - \mathbf{k} \cdot \mathbf{x}],$$

with $a^\mu = $ constant, satisfy the equations for the electromagnetic fields in the absence of sources. How are ω and $\mathbf{k}$ related?

(b) Defining $k^\mu = (k^0, \mathbf{k})$, $k^0 = \omega/c$, what is the condition on a^μ which ensures that these potentials are in Lorenz gauge? Show that it is also possible to impose the *radiation gauge* condition $\nabla \cdot \mathbf{A} = 0$. If $\boldsymbol{\epsilon}$ is a unit three-vector parallel to $\mathbf{a}$, show that we then have $\boldsymbol{\epsilon} \cdot \mathbf{k} = 0$.

(c) Obtain expressions for the electric and magnetic fields $\mathbf{E}$ and $\mathbf{B}$, and show that they oscillate with angular frequency ω and amplitudes $E = a\omega$ and $B = ak = a\omega/c$ respectively.

(d) Why are the electromagnetic waves so described called *waves*, why are they called *plane* waves, and what is the direction of their propagation?

(e) Take $\mathbf{k} = (0, 0, k)$ and $\boldsymbol{\epsilon} = (1, 0, 0)$. Write in matrix form the *field tensor* $F^{\mu\nu} = \partial^\mu A^\nu - \partial^\nu A^\mu$ and the *symmetrical stress tensor*

$$\Theta^{\mu\nu} = -\mu_0^{-1} \left(F^{\alpha\mu} F_\alpha{}^\nu - \frac{1}{4} \eta^{\mu\nu} F^{\alpha\beta} F_{\alpha\beta} \right).$$

3 This exercise uses the Lorentz transformation to explore the force on a moving charge exerted by a current-carrying wire.

Consider a long thin straight wire at rest in a frame K. We may take it to lie along the x-axis. We wish to determine the force on a charge q moving with velocity $\mathbf{v}$. The current in the wire, I say, is carried by the electrons which move with a (small!) velocity relative to the positive ions; we can suppose that the positive ions are at rest in K, while the conduction electrons have a drift velocity $\mathbf{u} = u\mathbf{i}$ along the wire. Let the linear charge density of the positive ions be λ_p, and of the conduction electrons be λ_n. Then

$$\lambda_p + \lambda_n = 0.$$

(a) Why?

(b) Express the current I carried by the wire in terms of u and λ.

In the laboratory frame K let the moving charge q be instantaneously at a point in the Oxy-plane distance r from the wire.

(c) What is the force (in the frame K) on q exerted by the *positive ions*.

The distance r of the charge q from the wire is the same in K as in the frame K' which is moving relative to K with velocity $\mathbf{u}$.

(d) Why?

In the frame K' the conduction electrons are at rest.

(e) What *in this frame* is the force on the charge q exerted by the conduction electrons?

(f) How is the linear charge density of the conduction electrons λ'_n in K' related to λ_n?

The Lorentz transformation for the components of a force on a particle moving with velocity $\mathbf{v}$ is given by

$$F'_x = F_x - \frac{(\beta/c)(v_y F_y + v_z F_z)}{[1 - (\beta/c)v_x]},$$

$$F'_y = \frac{F_y}{\gamma[1 - (\beta/c)v_x]},$$

$$F'_z = \frac{F_z}{\gamma[1 - (\beta/c)v_x]}.$$

(g)   Show that the force on the charge q in the frame K due to the conduction electrons can be written

$$\mathbf{F}_n = \frac{q}{2\pi\epsilon_0 r}\left[\frac{I}{c^2}v_y\mathbf{i} + \left(\lambda_n - \frac{v_x I}{c^2}\right)\mathbf{j}\right].$$

(h)   Adding to this the force in K due to the ions, show that the total force on q is

$$\mathbf{F} = \frac{\mu_0 q}{2\pi}\frac{I}{r}\mathbf{v} \times \mathbf{k},$$

and relate this to the Lorentz force $q\mathbf{v} \times \mathbf{B}$. Does $\mathbf{B}$ agree with what you would expect?

   The conduction electrons have a drift velocity which is typically $u \simeq 10^{-4}\mathrm{m} \cdot \mathrm{s}^{-1}$, so that γ differs from 1 by only about 1 part in 10^{25}, and therefore the two contributions to the total force (which have opposite signs) differ in magnitude only by 1 part in 10^{25}. It is remarkable, but true, that it is this tiny difference which we observe as the magnetic force exerted by a current-carrying wire on a moving charge!

Chapter 5

Lagrangian and Hamiltonian

In this chapter we will construct the Hamiltonian and Lagrangian functions for the electromagnetic field, and show how the field equations may be obtained from the action principle. But first we review Lagrange's method for the dynamics of particles.

5.1 Lagrange's Equations

Joseph-Louis Lagrange (1736–1813) used the methods of the calculus of variations to obtain a powerful and general formulation of mechanics.[1] The configuration of a dynamical system is to be specified by a set of coordinates $q^i(t)$, with $\dot{q}^i(t)$ the instantaneous values for their rate of change with respect to the time t. Then the *action* for a motion in which the system goes from $q^i(t_1), \dot{q}^i(t_1)$ to $q^i(t_2), \dot{q}^i(t_2)$ in the time interval (t_1, t_2) is given by

$$S(t_1, t_2) = \int_{t_1}^{t_2} L[(q^i(t), \dot{q}^i(t)] \, dt, \qquad (5.1)$$

where the Lagragian L is a function of the coordinates $q^i(t)$ and the velocities $\dot{q}^i(t)$. The q^i, $i = 1, \ldots, N$ can be *generalized coordinates*, not necessarily the Cartesian coordinates of the particles that comprise the system, but any set of N quantities that serve to specify the instantaneous configuration. Indeed the advantage of Lagrange's methods was to simplify the application of Newton's laws of motion to complicated systems, for example ones containing rigid bodies, because the internal forces that maintain their shape or other geometrical constraints on the system drop out from the analysis. It is assumed that the forces to be considered are *conservative*, so that they may be derived from a

---

[1] In his *Mécanique Analytique* (Desaint, Paris, 1788); reissued by Cambridge University Press, Cambridge, 2010.

potential $V(q^i)$. The kinetic energy T of the system may be expressed as a function of the $\dot{q}^i$ (and possibly also of the q^i). The Lagrangian is then

$$L = T - V, \tag{5.2}$$

and the action is the path integral

$$S = \int_{t_1}^{t_2} (T - V)\, dt. \tag{5.3}$$

Lagrange's equations follow from the *principle of stationary action*. The motion of the dynamical system is such as to make the action stationary under variations of the functions $q^i(t)$; explicitly then, on varying the path $q^i(t)$ with δq^i continuous but arbitrary for $t_1 < t < t_2$ but vanishing at $t = t_1$ and $t = t_2$, we obtain

$$\delta S = 0$$

$$= \int_{t_1}^{t_2} \sum_i \left[\frac{\partial L}{\partial \dot{q}^i} \delta \dot{q}^i + \frac{\partial L}{\partial q^i} \delta q^i \right] dt$$

$$= \int_{t_1}^{t_2} \sum_i \left[p_i \delta \dot{q}^i + \frac{\partial L}{\partial q^i} \delta q^i \right] dt$$

$$= \int_{t_1}^{t_2} \sum_i \left[\frac{d}{dt}(p_i \delta q^i) - (\dot{p}_i - \frac{\partial L}{\partial q^i}) \delta q^i \right] dt. \tag{5.4}$$

We have defined the *canonical momenta* p_i conjugate to the coordinates q^i by

$$p_i \equiv \frac{\partial L}{\partial \dot{q}^i}, \tag{5.5}$$

and with $\delta \dot{q}^i = \frac{d}{dt}(\delta q^i)$, on integration by parts we may set to zero $\int_{t_1}^{t_2} \sum_i [\frac{d}{dt}(p_i \delta q^i)]\, dt$ because the variation $\delta q^i(t)$ vanishes at the end points. There follows

$$\int_{t_1}^{t_2} \sum_i \left(\dot{p}_i - \frac{\partial L}{\partial q^i} \right) \delta q^i\, dt,$$

and since $\delta q^i(t)$ is arbitrary for $t_1 < t < t_2$, it follows that

$$\dot{p}_i \equiv \frac{\partial L}{\partial q^i} = F_i, \tag{5.6}$$

F_i being the generalized force consistent with the conservation of energy that requires $\delta W = \sum_i F_i \delta q^i$. So we have Lagrange's equations:

$$\frac{d}{dt} p_i = \frac{\partial L}{\partial q^i},$$

$$p_i = \frac{\partial L}{\partial \dot{q}^i}. \tag{5.7}$$

5.2 The Lagrangian for a Charged Particle

Consider as an example a nonrelativistic particle with kinetic energy

$$T = \frac{1}{2}m\mathbf{u}^2$$

with a potential $V(\mathbf{r})$. The Lagrangian is

$$L = T - V = \frac{1}{2}m\mathbf{u}^2 - V(\mathbf{r}). \tag{5.8}$$

We can think of this as being defined for *any* path $\mathbf{r}(t)$ with $\mathbf{u}(t) = \dot{\mathbf{r}}(t)$,

$$L = L[\mathbf{r}(t), \dot{\mathbf{r}}(t)],$$

and then may define the action for any path $\mathbf{r}(t)$ connecting some initial point $\mathbf{r}_1 = \mathbf{r}(t_1)$ to some final point $\mathbf{r}_2 = \mathbf{r}(t_2)$;

$$S = \int_{t_1}^{t_2} L[\mathbf{r}(t), \dot{\mathbf{r}}(t)]\, dt. \tag{5.9}$$

The action principle states that the action is stationary under variations of the path about the *actual* path followed by the particle in its motion from $\mathbf{r}_1$ at t_1 to $\mathbf{r}_2$ at t_2. Recapitulating the derivation in the preceding section, we have

$$\begin{aligned}
\delta S &= \int_{t_1}^{t_2} \left[\frac{\partial L}{\partial \mathbf{r}} \cdot \delta\mathbf{r} + \frac{\partial L}{\partial \dot{\mathbf{r}}} \cdot \delta\dot{\mathbf{r}} \right] dt \\
&= \int_{t_1}^{t_2} \left[\frac{\partial L}{\partial \mathbf{r}} \cdot \delta\mathbf{r} + \frac{d}{dt}\left(\frac{\partial L}{\partial \dot{\mathbf{r}}} \cdot \delta\mathbf{r} \right) - \frac{d}{dt}\left(\frac{\partial L}{\partial \dot{\mathbf{r}}} \right) \cdot \delta\mathbf{r} \right] dt \\
&= \int_{t_1}^{t_2} \left[\frac{\partial L}{\partial \mathbf{r}} - \frac{d}{dt}\left(\frac{\partial L}{\partial \dot{\mathbf{r}}} \right) \right] \cdot \delta\mathbf{r}\, dt, \tag{5.10}
\end{aligned}$$

where at the last step we use $\delta\mathbf{r} = 0$ at the end points. The vanishing of the integral for arbitrary $\delta\mathbf{r}$ satisfying the end-point condition is now equivalent to the statement of the Lagrange equations

$$\frac{d}{dt}\mathbf{p} = \frac{\partial L}{\partial \mathbf{r}},$$

where

$$\mathbf{p} = \frac{\partial L}{\partial \dot{\mathbf{r}}}.$$

These are indeed Newton's equations for the case considered, since $\mathbf{p}$ is the momentum and $\partial L/\partial\mathbf{r} = -\nabla V$ is the force.

We now seek a relativistic generalization. Since the condition $\delta S = 0$ must determine the same trajectory independent of the reference frame, we require that S should be a Lorentz scalar. But if

$$\begin{aligned}
S &= \int_{t_1}^{t_2} L\, dt \\
&= \int_{\tau_1}^{\tau_2} L\frac{dt}{d\tau}\, d\tau \tag{5.11}
\end{aligned}$$

is to be a scalar, it follows that $L\frac{dt}{d\tau} = \gamma L$ must be a scalar. For a *free* particle this must furthermore be independent of the position. But the only Lorentz scalar one can make out of the velocity alone is $U^\alpha U_\alpha = c^2$. So

$$\gamma L_{\text{free}} = \text{const.}$$

To get the correct nonrelativistic limit, this constant has to be $-mc^2$, and then

$$L_{\text{free}} = \frac{-mc^2}{\gamma(\mathbf{u})} = -mc^2\sqrt{1 - \frac{\mathbf{u}^2}{c^2}}. \tag{5.12}$$

This indeed gives

$$\mathbf{p} = \frac{\partial L_{\text{free}}}{\partial \mathbf{u}} = m\mathbf{u}\gamma(u)$$

and the equation of motion $\dot{\mathbf{p}} = 0$.

For a slowly moving charged particle, the interaction with the electromagnetic field introduces a term $V = q\Phi = qcA^0$ to the potential, and so leads to

$$L_{\text{int}}^{\text{NR}} = -qcA^0$$

as the expression for the nonrelativistic approximation to the interaction part of the Lagrangian. To go to the relativistic case, we try to find a Lorentz scalar to which this is an approximation, and then write $\gamma L_{\text{int}} =$ that Lorentz scalar. It is not hard to see that this has to be $-qU_\mu A^\mu$, so that we are led to consider

$$\gamma L = -mc^2 - qU_\mu A^\mu, \tag{5.13}$$

or

$$L = -mc^2\sqrt{1 - \frac{\mathbf{u}^2}{c^2}} - q(cA^0 - \mathbf{u}\cdot\mathbf{A}).$$

The momentum *canonically conjugate* to $\mathbf{r}$ is

$$\mathbf{P} = \frac{\partial L}{\partial \mathbf{u}} = m\gamma\mathbf{u} + q\mathbf{A}. \tag{5.14}$$

So in terms of the *mechanical* momentum $\mathbf{p}$ we have[2]

$$\mathbf{P} = \mathbf{p} + q\mathbf{A}. \tag{5.15}$$

One may then check that the Lagrange equations

$$\frac{d}{dt}P^i = \frac{\partial L}{\partial r^i} \tag{5.16}$$

do in fact give the correct equations of motion:

$$\frac{d}{dt}P^i = -qc\frac{\partial A^0}{\partial r^i} + q\sum_j u^j\frac{\partial A^j}{\partial r^i},$$

---

[2]Maxwell was prescient in calling $\mathbf{A}$ the electromagnetic momentum.

so that

$$\dot{p}^i + q\left(\frac{\partial A^i}{\partial t} + \sum_j \frac{\partial A^i}{\partial r^j}\frac{\partial r^j}{\partial t}\right) = -q\frac{\partial \Phi}{\partial r^i} + q\sum_j u^j \frac{\partial A^j}{\partial r^i},$$

which gives

$$\dot{\mathbf{p}} = q[\mathbf{E} + \mathbf{u} \times (\nabla \times \mathbf{A})]$$
$$= q[\mathbf{E} + \mathbf{u} \times \mathbf{B}]. \tag{5.17}$$

This is indeed what follows from the Lorentz force, and we also have

$$\frac{dU^\alpha}{d\tau} = \frac{q}{m}F^{\alpha\beta}U_\alpha. \tag{5.18}$$

5.3   The Hamiltonian for a Charged Particle

In general, the *Hamiltonian H* is defined by

$$H = \sum_i P_i \dot{q}^i - L(q^i, \dot{q}^i),$$

where P_i is the canonical momentum conjugate to q^i defined by

$$P_i \equiv \frac{\partial L}{\partial \dot{q}^i}.$$

The apparent dependence on $\dot{q}^i$ is illusory:

$$\frac{\partial H}{\partial \dot{q}^i} = P_i - \frac{\partial L}{\partial \dot{q}^i} = 0. \tag{5.19}$$

This confirms that H is a function only of the coordinates q^i and the canonically conjugate momenta P_i, So we may write

$$H = H(P_i, q^i).$$

And since $\frac{\partial L}{\partial q^i} = \dot{P}_i$ we have Hamilton's equations:

$$\frac{\partial H}{\partial P_i} = \dot{q}^i; \qquad \frac{\partial H}{\partial q^i} = -\dot{P}_i. \tag{5.20}$$

So starting with the Lagrangian for a relativistic charged particle found in the previous section, the Hamiltonian is defined by

$$H = \mathbf{P} \cdot \mathbf{u} - L, \tag{5.21}$$

in which $\mathbf{u}$ has to be eliminated in favor of $\mathbf{P}$. This leads to

$$H = (\mathbf{P} - q\mathbf{A}) \cdot \mathbf{u} + mc^2/\gamma + qcA^0$$

with
$$\mathbf{u} = \frac{\mathbf{P} - q\mathbf{A}}{m\gamma}$$

and
$$\gamma = \left[\frac{m^2 c^2 + (\mathbf{P} - q\mathbf{A})^2}{m^2 c^2} \right]^{1/2}$$

from which follows
$$H = [(\mathbf{P} - q\mathbf{A})^2 c^2 + m^2 c^4]^{\frac{1}{2}} + qcA^0$$
$$= [\mathbf{p}^2 c^2 + m^2 c^4]^{\frac{1}{2}} + qcA^0. \qquad (5.22)$$

The introduction of the electromagnetic field thus leads to the changes
$$\mathbf{p} \rightarrow \mathbf{P} = \mathbf{p} + q\mathbf{A},$$
$$p^0 = H/c \rightarrow P^0 = p^0 + qA^0. \qquad (5.23)$$

5.4   The Lagrangian for the Electromagnetic Field

Our first problem will be to understand how Lagrange's method may be further generalized to the situation of a *field* theory in which one has a continuous infinity of degrees of freedom (one—or a finite number—for every point in space). A way to approach this problem is suggested by a familiar model of a solid, in which we think of a crystal lattice of atoms (with lattice spacing a that will later be taken to zero), interacting with their nearest neighbors through some sort of elastic force, and for good measure also acted on by some other potential. For simplicity this can be taken to start with in just one dimension. If we suppose that the displacement of the ith atom is q^i, the potential energy, kinetic energy, and Lagrangian are then

$$V = \sum_i \left[\frac{1}{2} k(q^{i+1} - q^i)^2 + v(q^i) \right]$$
$$= \sum_i \left[\frac{1}{2} ka^2 \left(\frac{\Delta q^i}{a} \right)^2 + v(q^i) \right], \qquad (5.24)$$
$$T = \sum_i \frac{1}{2} m(\dot{q}^i)^2, \qquad (5.25)$$
$$L = T - V$$
$$\rightarrow \int ds \left[\frac{1}{2} \frac{m}{a} \left(\frac{\partial q(s,t)}{\partial t} \right)^2 - \frac{1}{2} ka \left(\frac{\partial q(s,t)}{\partial s} \right)^2 - \frac{1}{a} v(q(s,t)) \right]. \qquad (5.26)$$

At the last step we have suggested how to go to a continuum limit, where now we let $a \rightarrow 0$, keeping $\frac{m}{a}$, ka, and $\frac{v}{a}$ finite as the discrete index i is replaced by the continuous argument s of a field $q(s,t)$. This is in fact the appropriate way to model a continuous elastic medium. It also suggests that for a field theory we take as the Lagrangian an integral[3] of what is called the *Lagrangian density*,

$$L = \int d^3x\, \mathcal{L}\left[q(\mathbf{x},t), \boldsymbol{\nabla}q(\mathbf{x},t), \frac{\partial q(\mathbf{x},t)}{\partial t}\right]. \tag{5.27}$$

We then have for the action

$$S = \int dt \int d^3x\, \mathcal{L} \tag{5.28}$$

and the principle of stationary action $\delta S = 0$ gives the *Lagrange field equation*:

$$\frac{d}{dt}\left(\frac{\partial \mathcal{L}}{\partial(\frac{\partial q(\mathbf{x},t)}{\partial t})}\right) = \frac{\partial \mathcal{L}}{\partial q(\mathbf{x},t)} - \boldsymbol{\nabla}\left(\frac{\partial \mathcal{L}}{\partial \boldsymbol{\nabla}q(\mathbf{x},t)}\right),$$

which we can rewrite in a suggestive fashion by noting that since $\partial_\mu = (\frac{1}{c}\frac{\partial}{\partial t}, \boldsymbol{\nabla})$, with $\mathcal{L} = \mathcal{L}(q(\mathbf{x},t), \partial_\mu q(\mathbf{x},t))$ the Lagrange equation becomes

$$\partial_\mu\left[\frac{\partial \mathcal{L}}{\partial(\partial_\mu q(\mathbf{x},t))}\right] = \frac{\partial \mathcal{L}}{\partial q}. \tag{5.29}$$

This then is the *field equation*. For example, if

$$\mathcal{L} = \frac{1}{2}(\partial^\mu q)(\partial_\mu q) - \frac{\kappa^2}{2}q^2,$$

as might be suggested by our model of an elastic medium, the field equation is

$$\partial^\mu(\partial_\mu q) \equiv \Box q = -\kappa^2 q,$$

which is the *Klein-Gordon equation*. Note that if q is a Lorentz scalar, $\mathcal{L}$ is also a Lorentz scalar, and the resulting field equation is Lorentz invariant. Conversely , to get a Lorentz-invariant field equation, we must start from a Lorentz-invariant *action*, and this means in turn that $\mathcal{L}$ must be a Lorentz scalar.

Let us turn now to the electromagnetic field. The field variable will then not be the scalar field q of our previous example, but will be the four-vector A^α. The Lagrangian density depends on the derivatives of the fields, so instead of $\partial^\mu q$ it will involve $\partial^\beta A^\alpha$. In the previous example we had introduced $(\partial^\mu q)(\partial_\mu q)$ as a scalar, quadratic in the field gradients. On general grounds we expect that there should be a term in the Lagrangian of this form, and so are led to seek a Lorentz scalar quadratic in the derivatives $\partial^\beta A^\alpha$. Just as a Lorentz scalar Lagrangian density leads to Lorentz invariance of the field equations, so also *gauge* invariance of the Lagrangian density will lead to gauge invariance of the field equations. So we need to find a gauge-invariant Lorentz scalar quadratic in the field derivatives. We have already encountered two such, namely, $F^{\alpha\beta}F_{\alpha\beta}$

---

[3]The range of this integral, and others like it, will be over all space unless otherwise specified. We will assume that the integrand falls off sufficiently fast at infinity that the integral converges.

and $F^{\alpha\beta} * F_{\alpha\beta}$; but the latter of these is in fact only a *pseudo*scalar (it changes sign under reflections). This motivates the choice

$$\mathcal{L} = \text{const } F^{\alpha\beta} F_{\alpha\beta}$$

for the electromagnetic field in the absence of sources j^μ. If sources are present we expect to have to subtract from this the potential energy of the interaction between the sources and the electromagnetic field, so have

$$\mathcal{L} = \text{const } F^{\alpha\beta} F_{\alpha\beta} - j^\mu A_\mu \tag{5.30}$$

(the interaction term being just the generalization to a general source of what we have already discussed for the interaction of a single charged particle with the electromagnetic field) wherein $F_{\alpha\beta} \equiv \partial_\alpha A_\beta - \partial_\beta A_\alpha$. It might be objected that the interaction term is *not* gauge invariant; indeed under a gauge transformation it changes by $j^\mu \partial_\mu \chi$. But this differs from the total divergence $\partial_\mu(j^\mu \chi)$ (which makes no contribution to the field equations anyway, and so can be dropped) by $\chi \partial_\mu j^\mu$, which vanishes *because the current is conserved*. There is thus illustrated the very important and deep connection between the gauge invariance of the theory and the conservation law.[4]

The Lagrange equations which follow from the choice $\mathcal{L} = k F^{\alpha\beta} F_{\alpha\beta} - j^\mu A_\mu$ are

$$\partial_\beta \left(\frac{\partial \mathcal{L}}{\partial A_{\alpha,\beta}} \right) = \frac{\partial \mathcal{L}}{\partial A_\alpha} = -j^\alpha.$$

(We have introduced the very convenient notation $f_{,\beta}$ for $\partial_\beta f = \frac{\partial f}{\partial x^\beta}$, so that $A_{\alpha,\beta}$ means $\partial_\beta A_\alpha = \frac{\partial A_\alpha}{\partial x^\beta}$.) But

$$\frac{\partial \mathcal{L}}{\partial A_{\alpha,\beta}} = k \frac{\partial}{\partial A_{\alpha,\beta}} (F_{\rho\sigma} \eta^{\rho\mu} \eta^{\sigma\nu} F_{\mu\nu})$$

$$= 2k F^{\mu\nu} \frac{\partial}{\partial A_{\alpha,\beta}} (A_{\nu,\mu} - A_{\mu,\nu})$$

$$= 2k F^{\mu\nu} (\delta^\alpha_\nu \delta^\beta_\mu - \delta^\alpha_\mu \delta^\beta_\nu)$$

$$= -4k F^{\alpha\beta}, \tag{5.31}$$

so we have $-4k F^{\alpha\beta}{}_{,\beta} = -j^\alpha$, which is to be compared with the field equation

$$F^{\alpha\beta}{}_{,\beta} = -\mu_0 j^\alpha. \tag{5.32}$$

This fixes the normalization constant $k = -\frac{1}{4\mu_0}$. The Lagrangian density can also be written as

$$\mathcal{L} = -\frac{1}{4\mu_0} F^{\alpha\beta} F_{\alpha\beta} - j^\mu A_\mu$$

$$= -\frac{1}{2\mu_0} \left(\mathbf{B}^2 - \frac{\mathbf{E}^2}{c^2} \right) - j^\mu A_\mu$$

$$= \frac{1}{2} \left(\epsilon_0 \mathbf{E}^2 - \frac{1}{\mu_0} \mathbf{B}^2 \right) - j^\mu A_\mu. \tag{5.33}$$

---

[4]See Section 5.6 below.

5.5   The Hamiltonian for the Electromagnetic Field

The Hamiltonian (which generalizes $H = \sum_i p_i \dot{q}^i - L$) is

$$H = \int d^3x \, \frac{\partial \mathcal{L}}{\partial \dot{A}^\mu} \dot{A}^\mu - L \qquad (5.34)$$

since the momentum conjugate to A^μ is $\frac{\partial \mathcal{L}}{\partial \dot{A}^\mu}$. Just as was done in Section 5.3, we should eliminate the apparent dependence on the velocities (now the time derivatives of the A^μ) in favor of the momenta, and then find that the field equations that followed from Lagrange's approach may be obtained instead from the Hamiltonian equations of motion.

Note that since the time derivative of the component A^0 does not enter into the Lagrangian, the momentum conjugate to A^0 vanishes identically; this is a reflection of the fact that not all of the four components of A^μ are independent, and only three need be determined by the equations of motion, with A^0 being given from the gauge condition, for example $\partial_\mu A^\mu = 0$. The momentum conjugate to A^i, (where $i = 1, 2, 3$) is $\frac{\partial \mathcal{L}}{\partial \dot{A}^i} = -\epsilon_0 E^i$, and we then find

$$
\begin{aligned}
H &= -\int d^3x \, \epsilon_0 \mathbf{E} \cdot \dot{\mathbf{A}} - L \\
&= \int d^3x \, \left\{ \epsilon_0 [\mathbf{E} \cdot (\mathbf{E} + \boldsymbol{\nabla}\Phi)] - \frac{1}{2}\epsilon_0 [\mathbf{E}^2 - c^2 \mathbf{B}^2] + \rho\Phi - \mathbf{A} \cdot \mathbf{j} \right\} \\
&= \int d^3x \, \left\{ \frac{1}{2}\epsilon_0 [\mathbf{E}^2 + c^2 \mathbf{B}^2] - \mathbf{A} \cdot \mathbf{j} - \Phi[\boldsymbol{\nabla} \cdot (\epsilon_0 \mathbf{E}) - \rho] \right\}. \qquad (5.35)
\end{aligned}
$$

At the last step, we have dropped a term $\boldsymbol{\nabla} \cdot (\epsilon_0 \mathbf{E}\Phi)$ which is a divergence that makes no contribution to the (Hamilton) equations of motion. The only other place where Φ occurs is in the last remaining term, $-\Phi[\boldsymbol{\nabla} \cdot (\epsilon_0 \mathbf{E}) - \rho]$, and the corresponding Hamilton equation simply ensures the vanishing of the expression which multiplies Φ, namely, the Maxwell equation $\boldsymbol{\nabla} \cdot (\epsilon_0 \mathbf{E}) - \rho = 0$. If this equation is imposed as a *constraint*, Φ may be eliminated altogether, and we have

$$H = \int d^3x \, \left\{ \frac{1}{2}\epsilon_0 [\mathbf{E}^2 + c^2 \mathbf{B}^2] - \mathbf{A} \cdot \mathbf{j} \right\}.$$

In these remaining terms $\mathbf{B} = \boldsymbol{\nabla} \times \mathbf{A}$, and the field variable is $\mathbf{A}$, with conjugate momentum as previously given, $-\epsilon_0 \mathbf{E}$. So the Hamiltonian has been expressed entirely in terms of the fields and the conjugate momenta as is required. Hamilton's equations, the analogs of $\dot{p} = -\frac{\partial H}{\partial q}$, $\dot{q} = \frac{\partial H}{\partial p}$, are

$$
\begin{aligned}
-\epsilon_0 \dot{\mathbf{E}} &= \mathbf{j} - c^2 \epsilon_0 \boldsymbol{\nabla} \times \mathbf{B}, \\
\dot{\mathbf{A}} &= -\mathbf{E} - \boldsymbol{\nabla}\Phi. \qquad (5.36)
\end{aligned}
$$

And these are again Maxwell's equations. [You might notice that the constraint equation $\boldsymbol{\nabla} \cdot (\epsilon_0 \mathbf{E}) - \rho = 0$ is consistent with the first of these because the sources satisfy the conservation equation $\dot{\rho} + \boldsymbol{\nabla} \cdot \mathbf{j} = 0$, as we have already remarked.]

5.6   Noether's Theorem

We have emphasized the connection between the gauge invariance of the theory and the conservation of the electric current. This is an illustration of a general connection between symmetries and conserved quantities that was proved by the mathematician *Emmy Noether* (1882–1935) in 1915.[5] Her theorems have wide application, especially in the gauge symmetries of the relativistic quantum fields of the standard model of particle physics; but we will restrict ourselves to the simple case of the gauge invariance of electrodyanmics.

The invariance of the action

$$S = \int dt \int d^3x \, \mathcal{L}(A_\mu, A_{\mu,\nu})$$

$$= \int dt \int d^3x \left[-\frac{1}{4\mu_0} F^{\mu\nu} F_{\mu\nu} - j^\mu A_\mu \right] \tag{5.37}$$

under the gauge transformation

$$A_\mu \longrightarrow A_\mu - \partial_\mu \chi \tag{5.38}$$

(which leaves $F^{\mu\nu}$ invariant) is an example of a symmetry of the action under a continuous group of transformations generated by infintesimal changes in the function χ. Thus if $\chi \longrightarrow \chi + \delta\chi$, and $S \longrightarrow S + \delta S$, we assert that $\delta S = 0$. But

$$\delta S = \int dt \int d^3x \, \delta \left[-\frac{1}{4\mu_0} F^{\mu\nu} F_{\mu\nu} - j^\mu A_\mu \right]$$

$$= \int dt \int d^3x \, \delta \left[-j^\mu A_\mu \right]$$

$$= \int dt \int d^3x \, \delta \left[j^\mu \partial_\mu \chi \right]$$

$$= \int dt \int d^3x \, \delta \left[\partial_\mu (j^\mu \chi) - \chi \partial_\mu j^\mu \right]$$

$$= \int dt \int d^3x \, \delta \left[-\chi \partial_\mu j^\mu \right], \tag{5.39}$$

where at the last step we have discarded the total derivative which gives zero on integration if the sources are confined to a finite region. Then if δS is to vanish for arbitrary $\delta\chi$, we require the conservation of the current (now an example of a *Noether current*)

$$\partial_\mu j^\mu = 0.$$

There is another application of Noether's theorem, closer to the way it is used in the standard model. We have introduced a Lagrange density

$$\mathcal{L} = \frac{1}{2}(\partial^\mu q)(\partial_\mu q) - \frac{\kappa^2}{2} q^2$$

---

[5] "Invariante Variationsprobleme," Nachr. D. König. Gesellsch. D. Wiss. zu Göttingen, Math-phys. Klasse **1918**, 235–257.

for a field with field equation $\Box q = -\kappa^2 q$. But now consider *two* such fields, so that

$$\mathcal{L} = \left(\frac{1}{2}(\partial^\mu q_1)(\partial_\mu q_1) - \frac{\kappa^2}{2}q_1^2\right) + \left(\frac{1}{2}(\partial^\mu q_2)(\partial_\mu q_2) - \frac{\kappa^2}{2}q_2^2\right), \tag{5.40}$$

and consider the changes

$$q_1 \longrightarrow q_1 \cos\theta - q_2 \sin\theta; \quad q_2 \longrightarrow q_1 \sin\theta + q_2 \cos\theta,$$

or for infinitesimal θ

$$\delta q_1 = -q_2 \delta\theta; \quad \delta q_2 = q_1 \delta\theta. \tag{5.41}$$

So long as the parameter $\delta\theta$ is constant, not a function of x, this "global" symmetry leaves $\mathcal{L}$ invariant. But we want to extend this symmetry to be a "local" symmetry, that is, with $\delta\theta = \delta\theta(x)$. The term $q_1^2 + q_2^2$ in the Lagrangian does indeed remain invariant. But since

$$\delta(\partial_\mu q_1) = \partial_\mu \delta q_1 = -\partial_\mu(q_2 \delta\theta) = -(\partial_\mu q_2)\delta\theta - q_2 \partial_\mu \delta\theta,$$

and similarly

$$\delta(\partial_\mu q_2) = (\partial_\mu q_1)\delta\theta + q_1 \partial_\mu \delta\theta,$$

the terms $q_1 \partial_\mu \delta\theta$, $-q_2 \partial_\mu \delta\theta$ spoil the invariance. To cancel them, we introduce the *gauge field* A_μ and the *covariant derivative* D_μ with

$$D_\mu q_1 = \partial_\mu q_1 - A_\mu q_2; \quad D_\mu q_2 = \partial_\mu q_2 + A_\mu q_1. \tag{5.42}$$

Then if $\delta A_\mu = -\partial_\mu \delta\theta$ we find that

$$\begin{aligned}
\delta D_\mu q_1 &= \delta\left(\partial_\mu q_1 \, A_\mu q_2\right) \\
&= (-(\partial_\mu q_2)\delta\theta - q_2 \partial_\mu \delta\theta) - (-\partial_\mu \delta\theta)\, q_2 - A_\mu q_1 \delta\theta \\
&= -D_\mu q_2 \delta\theta. \tag{5.43}
\end{aligned}$$

The Lagrangian has now become

$$\begin{aligned}
\mathcal{L} &= \frac{1}{2}\left[(D^\mu q_1)(D_\mu q_1) + (D^\mu q_2)(D_\mu q_2) - \kappa^2(q_1^2 + q_2^2)\right] \\
&= \frac{1}{2}\left[(\partial^\mu q_1 - A^\mu q_2)(\partial_\mu q_1 - A_\mu q_2) + (\partial^\mu q_2 + A^\mu q_1)(\partial_\mu q_2 + A_\mu q_1) - \kappa^2(q_1^2 + q_2^2)\right] \\
&= \frac{1}{2}\left[(\partial^\mu q_1)(\partial_\mu q_1) - 2(\partial^\mu q_1)(A_\mu q_2) + (A_\mu A^\mu q_2^2)(\partial^\mu q_2)(\partial_\mu q_2) + 2(\partial^\mu q_2)(A_\mu q_1) + (A_\mu \right. \\
&\quad \left. - \kappa^2(q_1^2 + q_2^2)\right]. \tag{5.44}
\end{aligned}$$

We now determine the change in $\mathcal{L}$ when A_μ changes, $\delta A_\mu = -\partial_\mu \delta\theta$, and find

$$\begin{aligned}
\delta\mathcal{L} &= \left[-(\partial^\mu q_1)q_2 + A^\mu q_2^2 + (\partial^\mu q_2)q_1 + A^\mu q_1^2\right]\delta A_\mu \\
&= \left[(\partial^\mu q_1)q_2 - A^\mu q_2^2 - (\partial^\mu q_2)q_1 - A^\mu q_1^2\right]\partial_\mu \delta\theta \\
&= \left[q_2 D^\mu q_1 - q_1 D^\mu q_2\right]\partial_\mu \delta\theta \\
&= -\theta\partial_\mu\left[q_2 D^\mu q_1 - q_1 D^\mu q_2\right] + \partial_\mu\left[(q_2 D^\mu q_1 - q_1 D^\mu q_2)\delta\theta\right]. \tag{5.45}
\end{aligned}$$

On integration of $\mathcal{L}$ the second term, which, being a divergence, makes no contribution to the action. But if the first term is likewise to give zero for arbitrary $\delta\theta$, we again conclude that there is a conserved current:

$$\partial_\mu (q_2 D^\mu q_1 - q_1 D^\mu q_2) = 0. \tag{5.46}$$

This current, $j^\mu = (q_2 D^\mu q_1 - q_1 D^\mu q_2)$, which couples to the gauge field A_μ is the Noether current.[6]

5.7 Exercises for Chapter 5

1 Suppose that inside a large box of volume V, the potentials are $\mathbf{A}(\mathbf{x}, t) = \mathbf{a}(t)\sin(\mathbf{k}\cdot\mathbf{x})$, with $\mathbf{k}\cdot\mathbf{a} = 0$, and $\Phi = 0$; and that they vanish outside the box. Show that the Lagrangian for the electromagnetic fields can then be written as

$$L = \frac{V\epsilon_0}{4}(\dot{\mathbf{a}}^2 - c^2 k^2 \mathbf{a}^2).$$

Obtain a similar expression for the Hamiltonian H.

2 A certain Lagrangian density $\mathcal{L}_\psi(\psi_\alpha, A_\mu)$ for a set of fields ψ_α is invariant under the transformations $\psi_\alpha \to \psi_\alpha + \delta\psi_\alpha$, and $A_\mu \to A_\mu + \delta A_\mu$.
(a) What are the field equations for the fields ψ_α?
(b) Assuming that these field equations are satisfied, show that $j^\mu = \partial\mathcal{L}_\psi/\partial A_\mu$ is a conserved current, that is, $\partial_\mu j^\mu = 0$.

3 Point masses m are fixed to the ends of a rigid rod of length l and negligible mass that rotates freely about its mid-point.
(a) Express the kinetic energy of the rod in terms of the spherical polar and azimuthal angles θ and ϕ that define its orientation, and their time derivatives.
(b) What is the Lagrangian for this simple system?
(c) And what is the Hamiltonian?
(d) Show that the Lagrangian is invariant under the transformations $\theta \to \theta + \delta\theta$, and $\phi \to \phi + \delta\phi$.
(e) What is the Noether current corresponding to this invariance?

---

[6]In quantum field theory, the quanta associated with the fields q_1, q_2 may be identified as charged partcles and their antiparticles. If we define a *complex* field q by $q = q_1 + iq_2$ it is apparent that our transformation $q_1 \to q_1 \cos\theta - q_2 \sin\theta$; $q_2 \to q_1 \sin\theta + q_2 \cos\theta$ (a rotation in the $q_1 - q_2$ plane) is equivalent to $q \to e^{i\theta}q$. The transformation may now be identified as the action of an element of the *Lie group* $U(1)$, which is the group of 1×1 *unitary* matrices. A 1×1 matrix is just a number, and unitary means that the elements g of the group satisfy $g^\dagger = g^{-1}$, where in general $g^\dagger$ denotes the *Hermitian conjugate* = complex conjugate of the transpose of g (so for a number just the same as complex conjugate). The group $U(1)$ may be generalized, for example to $U(2)$, the group of 2×2 unitary matrices; this group has a subgroup $SU(2)$ which is the group of 2×2 unitary matrices with determinant 1. Likewise $SU(3)$ is the group of 3×3 traceless unitary matrices with determinant 1. These groups, $U(1), SU(2), SU(3)$ are associated with the symmetries of the standard model of particle physics. The quanta of the corresponding gauge fields are the photon, the vector mesons of the weak interactions, and the gluons, which transmit forces between the fundamental particles.

Chapter 6

Stress, Energy, and Momentum

This chapter will introduce a tensor that brings together the energy, momentum, and stress in the electromagnetic field. Maxwell used ideas related to the theory of elastic media to express in mathematical language what Faraday had described as the repulsion between his lines of force. By analogy with the internal forces in an elastic medium, we have a stress in the electromagnetic field, the force that the field exerts across a surface separating any region of the field from neighboring regions or contiguous material bodies. In modern language, this is described through a tensor, the force (itself a three-vector) across an elementary area of the surface which has its orientation defined by the vector direction of its normal (another three-vector). So the stress tensor is, in three dimensions, a 3×3 component object. This had been described around 1822 in the context of elasticity theory by the French mathematician Augustin-Louis Cauchy (1789–1857), and considerations of equilibrium show it to be symmetrical.

6.1   The Canonical Stress Tensor

We have introduced the Lagrangian density

$$\mathcal{L} = -\frac{1}{4\mu_0} F^{\alpha\beta} F_{\alpha\beta} - j^{\mu} A_{\mu}$$

$$= \frac{1}{2}\epsilon_0 (\mathbf{E}^2 - c^2 \mathbf{B}^2) - j^{\mu} A_{\mu}, \tag{6.1}$$

which is a Lorentz scalar. This means that the field equations, that is, Maxwell's equations, which follow from the Lagrangian

$$L = \int \mathcal{L} \, d^3x \tag{6.2}$$

and the action

$$S = \int L \, dt = \frac{1}{c} \int \mathcal{L} \, d^4x \tag{6.3}$$

are Lorentz covariant—they have the same form in all inertial frames of reference.

We have defined the Hamiltonian H and so also the Hamiltonian density $\mathcal{H}$,

$$H = \int \mathcal{H}\, d^3x, \tag{6.4}$$

$$\mathcal{H} = \frac{\partial \mathcal{L}}{\partial A_{\mu,0}} A_{\mu,0} - \mathcal{L}$$
$$= \frac{1}{2}\epsilon_0(\mathbf{E}^2 + c^2\mathbf{B}^2) + \nabla \cdot (\epsilon_0 \Phi \mathbf{E}) - \mathbf{j} \cdot \mathbf{A}. \tag{6.5}$$

This does not behave in a nice way under Lorentz transformations. But it suggests the definition

$$T^\nu{}_\lambda = \frac{\partial \mathcal{L}}{\partial A_{\mu,\nu}} A_{\mu,\lambda} - \delta^\nu_\lambda \mathcal{L}$$
$$= \frac{1}{\mu_0} F^{\mu\nu} A_{\mu,\lambda} - \delta^\nu_\lambda \mathcal{L}, \tag{6.6}$$

which is a tensor of which $\mathcal{H}$ is a component. This tensor is called the *canonical stress tensor*. We have

$$T^{0\lambda} = (\mathcal{H}, \mathbf{\Pi}), \tag{6.7}$$

with $T^{00} = \mathcal{H}$ as given above, and[1]

$$\Pi^i = T^{0i} = \frac{1}{\mu_0 c}\left\{(\mathbf{E} \times \mathbf{B})^i + \nabla \cdot (\mathbf{E}A^i) - \frac{\rho}{\epsilon_0}A^i\right\}. \tag{6.8}$$

Let us recall that u, the energy density in the electromagnetic field, is

$$u = \frac{1}{2}(\mathbf{E} \cdot \mathbf{D} + \mathbf{B} \cdot \mathbf{H})$$
$$= \frac{1}{2}\epsilon_0(\mathbf{E}^2 + c^2\mathbf{B}^2). \tag{6.9}$$

We also define the *Poynting vector* $\mathbf{S}$:

$$\mathbf{S} = \mathbf{E} \times \mathbf{H}$$
$$= \frac{1}{\mu_0}\mathbf{E} \times \mathbf{B}. \tag{6.10}$$

This vector is the flux vector of electromagnetic energy as may be seen by recognizing that the equation

$$\dot{u} + \nabla \cdot \mathbf{S} = -\mathbf{j} \cdot \mathbf{E}, \tag{6.11}$$

which follows directly from Maxwell's equations, expresses the conservation of energy, since the right-hand side is the work done by the Lorentz force on the current. Then

$$\Pi^i = \frac{1}{c}S^i + \nabla \cdot (c\epsilon_0 \mathbf{E}A^i) - j^0 A^i, \tag{6.12}$$

$$\mathcal{H} = u + \nabla \cdot (c\epsilon_0 \mathbf{E}A^0) - j^0 A_0 + j^\mu A_\mu. \tag{6.13}$$

---

[1] Greek indices like λ have the range $0, 1, 2, 3$, while Roman indices like i have the range $1, 2, 3$.

6.2   The Symmetrical Stress Tensor

Suppose now that the electromagnetic fields are localized in some region, and that there are no sources. Then, integrating over that region,

$$\int T^{00}\, d^3x = \int [u + \boldsymbol{\nabla} \cdot (c\epsilon_0 \mathbf{E} A^0)]\, d^3x$$

$$= \int u\, d^3x$$

$$= U_{\text{field}}; \tag{6.14}$$

$$\int T^{0i}\, d^3x = \int \left[\frac{1}{c} S^i + \boldsymbol{\nabla} \cdot (c\epsilon_0 \mathbf{E} A^i) \right] d^3x$$

$$= \frac{1}{c} \int S^i\, d^3x$$

$$= cP^i_{\text{field}}. \tag{6.15}$$

Here U_{field} and $\mathbf{P}_{\text{field}}$ are respectively the total energy and momentum of the electromagnetic field in the region considered.

In the absence of sources, the energy conservation law is $\frac{\partial u}{\partial t} + \boldsymbol{\nabla} \cdot \mathbf{S} = 0$, which in spite of appearances is *not* in a covariant form, since u and $\mathbf{S}$ are not components of a four-vector. But it does indicate how we might find a covariant extension of it. We consider

$$\partial_\alpha T^{\alpha\beta} = \partial_\alpha \left[\frac{\partial \mathcal{L}}{\partial(\partial_\alpha A_\mu)} \partial^\beta A_\mu - \eta^{\alpha\beta} \mathcal{L} \right]$$

$$= \partial_\alpha \left[\frac{\partial \mathcal{L}}{\partial(\partial_\alpha A_\mu)} \right] \partial^\beta A_\mu + \frac{\partial \mathcal{L}}{\partial(\partial_\alpha A_\mu)} \partial_\alpha \partial^\beta A_\mu - \partial^\beta \mathcal{L}. \tag{6.16}$$

Now use the Lagrange equations

$$\partial_\alpha \frac{\partial \mathcal{L}}{\partial(\partial_\alpha A_\mu)} = \frac{\partial \mathcal{L}}{\partial A_\mu}$$

to obtain

$$\partial_\alpha T^{\alpha\beta} = \frac{\partial \mathcal{L}}{\partial A_\mu} \partial^\beta A_\mu + \frac{\partial \mathcal{L}}{\partial(\partial_\alpha A_\mu)} \partial^\beta (\partial_\alpha A_\mu) - \partial^\beta \mathcal{L},$$

which vanishes identically (use the chain rule for differentiation of $\mathcal{L} = \mathcal{L}(A_\mu, \partial_\alpha A_\mu)$). Thus the *covariant* conservation law

$$\partial_\alpha T^{\alpha\beta} = 0 \tag{6.17}$$

is a consequence of the equation of motion.

If this is integrated over the region containing the fields,

$$\int \partial_\alpha T^{\alpha\beta} = 0$$

$$= \int (\partial_0 T^{0\beta} + \partial_i T^{i\beta})\, d^3x$$

$$= \partial_0 \int T^{0\beta}\, d^3x + \text{surface term}. \tag{6.18}$$

The surface term vanishes, and what we have obtained is

$$\frac{d}{dt} U_{\text{field}} = 0, \tag{6.19}$$

$$\frac{d}{dt} \mathbf{P}_{\text{field}} = 0, \tag{6.20}$$

the conservation of the energy and of the momentum for localized fields in the absence of sources.

This is very pretty, but there are a number of objections:

• The electromagnetic energy and momentum ought to be covariantly defined as parts of a four-vector; we have implicitly been using a frame in which the observer is at rest.

• $\mathcal{H}$ and $\mathbf{\Pi}$ differ from u and $\mathbf{S}$ even in the absence of sources (albeit by a divergence).

• The tensor $T^{\alpha\beta}$ is not symmetrical. The significance of this arises from the wish to incorporate into a covariant conservation law the conservation of the *angular momentum* of the electromagnetic field. Thus

$$\mathbf{M}_{\text{field}} = \frac{1}{c} \int \mathbf{x} \times (\mathbf{E} \times \mathbf{H}) \, d^3x$$

$$= \frac{1}{c\mu_0} \int \mathbf{x} \times (\mathbf{E} \times \mathbf{B}) \, d^3x \tag{6.21}$$

is conserved, and a local covariant generalization of this would be

$$\partial_\alpha M^{\alpha\beta\gamma} = 0 \tag{6.22}$$

with

$$M^{\alpha\beta\gamma} = T^{\alpha\beta} x^\gamma - T^{\alpha\gamma} x^\beta.$$

This requires

$$\partial_\alpha (T^{\alpha\beta} x^\gamma - T^{\alpha\gamma} x^\beta) = 0,$$

that is,

$$(\partial_\alpha T^{\alpha\beta}) x^\gamma + T^{\alpha\beta} \delta^\gamma_\alpha - (\partial_\alpha T^{\alpha\gamma}) x^\beta - T^{\alpha\gamma} \delta^\beta_\alpha = 0,$$

or using the previous result

$$T^{\gamma\beta} - T^{\beta\gamma} = 0.$$

• On general grounds it is expected that $T^\alpha{}_\alpha = 0$. This is related to the fact that the quanta of the electromagnetic field, the photons, have zero mass, and likewise to the scale invariance of electromagnetism—these are technical points outside the scope of this book.

• If $T^{\alpha\beta}$ is to be of direct physical significance, it ought to be gauge invariant.

We seek to remedy these defects by modifying the tensor:

$$T^{\alpha\beta} \rightarrow \Theta^{\alpha\beta} = T^{\alpha\beta} - T_D^{\alpha\beta}$$

in such a way that

(a) $\quad \Theta^{\alpha\beta} = \Theta^{\beta\alpha}, \qquad \Theta$ is symmetric;

(b) $\quad \Theta^{\alpha}{}_{\alpha} = 0, \qquad \Theta$ is traceless;

(c) $\qquad \Theta$ is gauge invariant;

(d) $\quad \partial_{\alpha}\Theta^{\alpha\beta} = 0, \qquad \Theta$ is conserved;

(e) $\qquad \int \Theta^{0\beta} d^3x = \int T^{0\beta} d^3x.$

In the absence of sources, we had

$$T^{\nu}{}_{\lambda} = \frac{1}{\mu_0} F^{\mu\nu} A_{\mu,\lambda} - \delta^{\nu}_{\lambda}\mathcal{L}$$

$$= \frac{1}{\mu_0} [F^{\mu\nu}(A_{\mu,\lambda} - A_{\lambda,\mu}) + (F^{\mu\nu} A_{\lambda})_{,\mu} - F^{\mu\nu}{}_{,\mu}A_{\lambda}] - \delta^{\nu}_{\lambda}\mathcal{L}$$

$$= -\frac{1}{\mu_0} F^{\mu\nu} F_{\mu\lambda} - \delta^{\nu}_{\lambda}\mathcal{L} + \frac{1}{\mu_0}(F^{\mu\nu} A_{\lambda})_{,\mu}. \tag{6.23}$$

So let us define

$$T_D{}^{\nu}{}_{\lambda} = -\frac{1}{\mu_0}(F^{\mu\nu} A_{\lambda})_{,\mu}. \tag{6.24}$$

We then have

$$\Theta^{\nu\lambda} = -\frac{1}{\mu_0}\left(F^{\mu\nu} F^{\sigma\lambda}\eta_{\mu\sigma} - \frac{1}{4}\eta^{\nu\lambda}F^{\alpha\beta} F_{\alpha\beta}\right), \tag{6.25}$$

which is clearly

(a) symmetrical

(b) traceless (use $\delta^{\alpha}_{\alpha} = 4!$)

(c) gauge invariant, since it depends only on $F^{\mu\nu}$

(d) conserved in the absence of sources, since

$$T_D{}^{\nu}{}_{\lambda,\nu} = \partial_{\nu}\left(-\frac{1}{\mu_0}(F^{\mu\nu} A_{\lambda})_{,\mu}\right)$$

$$= -\frac{1}{\mu_0}(F^{\mu\nu} A_{\lambda})_{,\mu,\nu}, \tag{6.26}$$

which vanishes because $F^{\mu\nu}$ is antisymmetrical in μ, ν while the partial derivatives are symmetrical. Thus

$$\Theta^{\nu}{}_{\lambda,\nu} = T^{\nu}{}_{\lambda,\nu} = 0. \tag{6.27}$$

and

(e) for localized fields, the space integrals of $\Theta^{0\beta}$ and $T^{0\beta}$ are equal, since

$$
\begin{aligned}
\int \Theta^{0\beta} \, d^3x - \int T^{0\beta} \, d^3x &= \int T_D{}^{0\beta} \, d^3x \\
&= -\frac{1}{\mu_0} \int (F^{\mu 0} A^\beta)_{,\mu} \, d^3x \\
&= -\frac{1}{\mu_0} \int (F^{i0} A^\beta)_{,i} \, d^3x \\
&= -\frac{1}{\mu_0 c} \int \boldsymbol{\nabla} \cdot (\mathbf{E} A^\beta) \, d^3x,
\end{aligned}
\tag{6.28}
$$

which gives a surface integral that vanishes because $\mathbf{E} = 0$ on the boundary, the fields being localized.

We are thus motivated to define, even when there are sources present, the *symmetric stress tensor* $\Theta^{\alpha\beta}$ by

$$
\Theta^{\alpha\beta} \equiv -\frac{1}{\mu_0} \left[F^{\lambda\alpha} F_\lambda{}^\beta - \frac{1}{4} \eta^{\alpha\beta} F^{\mu\nu} F_{\mu\nu} \right],
\tag{6.29}
$$

for which

$$
\begin{aligned}
\Theta^{00} &= \frac{1}{2} \epsilon_0 (\mathbf{E}^2 + c^2 \mathbf{B}^2) = u, \\
\Theta^{0i} &= \frac{1}{c\mu_0} (\mathbf{E} \times \mathbf{B})^i = \frac{1}{c} S^i \equiv cg^i, \\
\Theta^{ij} &= -\epsilon_0 \left[E^i E^j + c^2 B^i B^j - \frac{1}{2} \delta^{ij} (\mathbf{E}^2 + c^2 \mathbf{B}^2) \right] \\
&= -(\text{the Maxwell stress tensor}).
\end{aligned}
\tag{6.30}
$$

This tensor thus combines the energy density u, the Poynting vector $\mathbf{S}$ (or the momentum density vector $\mathbf{g}$), and the Maxwell stress tensor (a *three*-tensor that gives the mechanical stress present in the electromagnetic field, which is responsible for the repulsion of lines of force described by Faraday) into a Lorentz covariant tensor.

6.3   The Conservation Laws with Sources

To see what happens to the conservation law in the presence of sources, consider

$$
\begin{aligned}
\partial_\alpha \Theta^{\alpha\beta} &= -\frac{1}{\mu_0} \left[F^{\lambda\alpha}{}_{,\alpha} F_\lambda{}^\beta + F^{\lambda\alpha} F_\lambda{}^\beta{}_{,\alpha} - \frac{1}{2} F^{\lambda\mu} F_{\lambda\mu,}{}^\beta \right] \\
&= j^\lambda F_\lambda{}^\beta - \frac{1}{2\mu_0} [2 F^{\lambda\alpha} F_\lambda{}^\beta{}_{,\alpha} - F^{\lambda\mu} F_{\lambda\mu,}{}^\beta].
\end{aligned}
\tag{6.31}
$$

Thus

$$\partial_\alpha \Theta^{\alpha\beta} - j^\lambda F_\lambda{}^\beta = -\frac{1}{2\mu_0}[2F_{\lambda\mu}F^{\lambda\beta,\mu} - F_{\lambda\mu}F^{\lambda\mu,\beta}]$$

$$= -\frac{1}{2\mu_0}F_{\lambda\mu}[F^{\lambda\beta,\mu} - F^{\mu\beta,\lambda} - F^{\lambda\mu,\beta}]$$

$$= \frac{1}{2\mu_0}F_{\lambda\mu}[F^{\beta\lambda,\mu} + F^{\mu\beta,\lambda} + F^{\lambda\mu,\beta}]$$

$$= 0, \tag{6.32}$$

where we make repeated use of the antisymmetry of $F_{\mu\nu}$ and at the last stage of Maxwell's homogeneous equation.

Thus we have derived

$$\partial_\alpha \Theta^{\alpha\beta} = j^\lambda F_\lambda{}^\beta \equiv -f^\beta, \tag{6.33}$$

having introduced the *Lorentz force density* $f^\beta = F^{\beta\lambda}j_\lambda = (f_0, \mathbf{f})$. We have

$$f^i = F^{i\lambda}j_\lambda = F^{i0}j_0 - \sum_k F^{ik}j^k = E^i\rho + (\mathbf{j} \times \mathbf{B})^i,$$

or

$$\mathbf{f} = \rho\mathbf{E} + \mathbf{j} \times \mathbf{B}; \tag{6.34}$$

and

$$f^0 = F^{0\lambda}j_\lambda = F^{0i}j_i = \left(-\frac{\mathbf{E}}{c}\right) \cdot (-\mathbf{j}) = \frac{\mathbf{E} \cdot \mathbf{j}}{c}. \tag{6.35}$$

These should be compared with

$$\mathbf{F} = q(\mathbf{E} + \mathbf{v} \times \mathbf{B})$$

and

$$\mathbf{F} \cdot \mathbf{v} = \mathbf{E} \cdot (q\mathbf{v});$$

respectively the Lorentz force (i.e., the rate of change of momentum) of a charged particle and the rate at which the field does work (i.e., the rate of change of the energy) on a charged particle. Thus the four-vector f^β gives the rate of change of the energy and the momentum of the *sources*. In other words

$$\int f^\beta \, d^3x = \frac{d}{dt}P^\beta_{\text{matter}}. \tag{6.36}$$

What this means is that

$$\int (\partial_\alpha \Theta^{\alpha\beta} + f^\beta) \, d^3x = 0$$

$$= \int (\partial_0 \Theta^{0\beta} + \partial_i \Theta^{i\beta}) \, d^3x + \frac{d}{dt}P^\beta_{\text{matter}}$$

$$= \int \partial_0 \Theta^{0\beta} \, d^3x + \int \Theta^{i\beta}n_i \, dS + \frac{d}{dt}P^\beta_{\text{matter}}$$

$$= \frac{d}{dt}\left[P^\beta_{\text{field}} + P^\beta_{\text{matter}}\right] + \text{surface term}, \tag{6.37}$$

where

$$P^{\beta}_{\text{field}} = \int \frac{1}{c} \Theta^{0\beta} \, d^3x$$

$$= \int \left(\frac{u}{c}, \mathbf{g} \right) d^3x$$

$$= \int \left(\frac{1}{c}(\text{energy density}), \text{momentum density} \right) d^3x, \qquad (6.38)$$

and the surface term vanishes for a closed system. At the differential level, the conservation equation gives (with $\beta = 0$)

$$\frac{1}{c}\left(\frac{\partial u}{\partial t} + \mathbf{\nabla} \cdot \mathbf{S} \right) = -\frac{\mathbf{E} \cdot \mathbf{j}}{c}, \qquad (6.39)$$

which is *Poynting's equation*; and (with $\beta = i$)

$$\frac{\partial g^i}{\partial t} = \left(T^{\text{Maxwell}} \right)^{ij}{}_{,j} - (\rho\mathbf{E} + \mathbf{j} \times \mathbf{B})^i, \qquad (6.40)$$

which expresses the fact that the rate of change of the density of momentum equals a contribution from the Maxwell stress exerted by the neighboring field and a term that is the equal and opposite reaction to the force exerted *by* the field *on* the sources.

6.4   The Field as an Ensemble of Oscillators

Although we have not yet discussed electromagnetic *radiation* in much detail, it is already possible to give an important and (I hope!) amusing result relating to radiation.

Consider an enclosure filled with electromagnetic radiation, but devoid of sources. The Hamiltonian is

$$H = \frac{1}{2}\epsilon_0 \int d^3x [\mathbf{E}^2 + c^2(\mathbf{\nabla} \times \mathbf{A})^2].$$

In a radiation gauge ($\text{div}\,\mathbf{A} = 0, \Phi = 0$), we have

$$H = \frac{1}{2}\epsilon_0 \int d^3x [\left(\frac{\partial \mathbf{A}}{\partial t} \right)^2 + c^2(\mathbf{\nabla} \times \mathbf{A})^2].$$

We may analyze the field into a superposition of *normal modes*:

$$\mathbf{A}(\mathbf{x}, t) = \sum_{\lambda} \frac{1}{\sqrt{\epsilon_0}} q_{\lambda}(t) \mathbf{A}_{\lambda}(\mathbf{x}).$$

The field equation $\Box\mathbf{A} = 0$ implies for the normal mode potentials

$$\nabla^2 \mathbf{A}_{\lambda} + \frac{\omega_{\lambda}^2}{c^2} \mathbf{A}_{\lambda} = 0,$$

while the amplitudes q_λ satisfy

$$\ddot{q}_\lambda(t) + \omega_\lambda^2 q_\lambda(t) = 0;$$

the normal mode frequencies ω_λ are as usual constants of separation of the t-from the $\mathbf{x}$-variables. The gauge choice implies that $\text{div}\mathbf{A}_\lambda = 0$, and the factor $\frac{1}{\sqrt{\epsilon_0}}$ in the normal mode expansion is chosen for later convenience. The functions $\mathbf{A}_\lambda(\mathbf{x})$ can be chosen to satisfy the orthonormality condition

$$\int d^3x \, \mathbf{A}_\lambda(\mathbf{x}) \cdot \mathbf{A}_\mu(\mathbf{x}) = \delta_{\lambda\mu}.$$

In terms of the normal modes, we have

$$
\begin{aligned}
H = {} & \frac{1}{2}\epsilon_0 \int d^3x \left[\left(\sum_\lambda \frac{1}{\sqrt{\epsilon_0}}\dot{q}_\lambda \mathbf{A}_\lambda\right) \cdot \left(\sum_\mu \frac{1}{\sqrt{\epsilon_0}}\dot{q}_\mu \mathbf{A}_\mu\right) \right. \\
& \left. + c^2 \left(\sum_\lambda \frac{1}{\sqrt{\epsilon_0}}q_\lambda \boldsymbol{\nabla} \times \mathbf{A}_\lambda\right) \cdot \left(\sum_\mu \frac{1}{\sqrt{\epsilon_0}}q_\mu \boldsymbol{\nabla} \times \mathbf{A}_\mu\right)\right].
\end{aligned}
$$

The last term may be manipulated as follows:

$$
\begin{aligned}
(\boldsymbol{\nabla} \times \mathbf{A}_\lambda) \cdot (\boldsymbol{\nabla} \times \mathbf{A}_\mu) = {} & \mathbf{A}_\lambda \cdot (\boldsymbol{\nabla} \times (\boldsymbol{\nabla} \times \mathbf{A}_\mu)) + \text{a divergence which integrates to} \\
= {} & \mathbf{A}_\lambda \cdot [\boldsymbol{\nabla}(\boldsymbol{\nabla} \cdot \mathbf{A}_\mu) - \nabla^2 \mathbf{A}_\mu] \\
= {} & \mathbf{A}_\lambda \cdot (-\nabla^2 \mathbf{A}_\mu) \quad \text{since div } \mathbf{A}_\mu = 0 \\
= {} & \mathbf{A}_\lambda \cdot \frac{\omega_\mu^2}{c^2} \mathbf{A}_\mu.
\end{aligned}
$$

Putting this into the previous expression for H gives $H = \frac{1}{2}\sum_\mu[\dot{q}_\mu^2 + \omega_\mu^2 q_\mu^2]$, or, better

$$H = \frac{1}{2}\sum_\mu[p_\mu^2 + \omega_\mu^2 q_\mu^2],$$

which is immediately recognized as the Hamiltonian for a system of simple harmonic oscillators. This result might have been known to Planck in 1900! ·

6.5 Exercises for Chapter 6

1 What are the components T^{ij} of the canonical stress tensor? How (if at all) do they differ from the components Θ^{ij} of the symmetrical stress tensor?

2 A steady current I is maintained in a long straight resistor of resistance R by a battery connected to the resistor by wires of negligible resistance.

(a) What is the electric field close to the resistor?

(b) What is the magnetic field close to the resistor?

(c) What is the Poynting vector close to the resistor?

(d) What is the rate at which energy is dissipated in the resistor in the form of heat?

(e) Where does this energy come from? How does it reach the resistor?

3   Consider the Lagrangian density

$$\mathcal{L}' = -\frac{1}{8\mu_0}\left(\partial_\alpha A_\beta\right)\left(\partial^\alpha A^\beta\right) - j^\mu A_\mu.$$

Does the action that follows from this Lagrangian density differ from that which follows from $\mathcal{L}$? Are the Lagrange field equations that follow from $\mathcal{L}'$ the same as those that follow from $\mathcal{L}$? What further assumptions (if any) are needed in order that $\mathcal{L}'$ could be used instead of $\mathcal{L}$ for the Lagrangian of Maxwell's theory?

Chapter 7

Motion of a Charged Particle

In this chapter we will derive some results relating to the motion of a charged particle in a given external field. But first let us consider a complementary problem, namely, the electromagnetic fields produced by a charged particle in unaccelerated motion.

7.1   Fields from an Unaccelerated Particle

Consider a particle with charge Q moving in frame K with constant velocity $\mathbf{v} = c\boldsymbol{\beta} = c\beta\mathbf{i}$, which has been chosen to be along the x-axis; if it passes through the origin at time $t = 0$, its position in the frame K is given by $\mathbf{v}t$. In the frame K', moving with this same velocity with respect to K, the particle is at rest at the origin. For a particle at rest the only field it generates is the familiar Coulomb electric field. So in the frame K' we have

$$\mathbf{E}' = \frac{q}{4\pi\epsilon_0} \frac{1}{r'^2} \frac{\mathbf{r}'}{r'}, \tag{7.1}$$

$$\mathbf{B}' = 0, \tag{7.2}$$

where $\mathbf{r}'$ is the position vector in frame K' of the field point P. In the frame K' it is the field point P that moves, with velocity $-\mathbf{v}$, so taking its position at $t' = 0$ to be on the y-axis, we have $\mathbf{r}' = (-c\beta t', b, 0)$. In summary, in frame K' at time t', the charge Q is at $(0,0,0)$, P is at $(-c\beta t', b, 0)$, and the vector from Q to P is $\mathbf{r}' = (-c\beta t', b, 0)$.

Now how do things look in the frame K? At time t, Q is at $(c\beta t, 0, 0)$, the field point P is at $(0, b, 0)$ and the vector from Q to P is $\mathbf{r} = (-c\beta t, b, 0)$. The fields at P may now be found by a Lorentz transformation of the fields in K', that is, the inverse of the transformation from K to K' given previously. Since we have $\mathbf{B}' = 0$, this gives

$$\mathbf{E} = \gamma\mathbf{E}' - \frac{\gamma^2}{\gamma+1}\frac{\mathbf{v}}{c}\left(\frac{\mathbf{v}}{c}\cdot\mathbf{E}'\right)$$

$$= \gamma\mathbf{E}' - (\gamma-1)\mathbf{i}(\mathbf{i}\cdot\mathbf{E}')$$

$$= \frac{q\gamma}{4\pi\epsilon_0}(b^2 + \gamma^2\beta^2 c^2 t^2)^{-\frac{3}{2}}(-\beta ct, b, 0); \tag{7.3}$$

$$\mathbf{B} = \frac{\gamma}{c}\frac{\mathbf{v}}{c}\times\mathbf{E}'$$

$$= \frac{q\gamma}{4\pi\epsilon_0}(b^2 + \gamma^2\beta^2 c^2 t^2)^{-\frac{3}{2}}\frac{\beta}{c^3}(0, 0, b). \tag{7.4}$$

(As usual, we have taken $\mathbf{i}, \mathbf{j}, \mathbf{k}$ as the unit vectors along, respectively, the x, y, z axes; and we have used $\frac{\beta^2\gamma^2}{\gamma+1} = \gamma - 1$.)

It is instructive to caculate the Poynting vector $\mathbf{S} = \frac{1}{\mu_0}\mathbf{E}\times\mathbf{B}$ in K (of course it vanishes in K'):

$$\mathbf{S} = \frac{q^2}{4\pi\epsilon_0}(b^2 + \gamma^2\beta^2 c^2 t^2)^{-3}\gamma^2\beta cb(b, vt, 0). \tag{7.5}$$

If now v is small, that is, $|\beta| \ll 1$, $\gamma \approx 1$,

$$\mathbf{E} \approx \frac{q}{4\pi\epsilon_0}\frac{1}{r^2}\frac{\mathbf{r}}{r}, \tag{7.6}$$

which is the static Coulomb field; and

$$\mathbf{B} \approx \frac{\mu_0}{4\pi}q\frac{\mathbf{v}\times\mathbf{r}}{r^3} = \frac{\mu_0}{4\pi}\frac{\mathbf{j}\times\mathbf{r}}{r^3}, \tag{7.7}$$

as expected from the Biot-Savart law.

If on the contrary $\beta \approx 1$, the observer at P in K experiences a short pulse of $\mathbf{E}/c$ and $\mathbf{B}$ of nearly equal intensity approximating to a short pulse of plane-polarized electromagnetic radiation propagating in the x-direction, and with $E^y = \frac{c}{\beta}B^z$ rising and then falling rapidly as the particle passes by, while E^x, which changes sign rapidly, has no effect on average.

7.2 Motion of a Particle in an External Field

7.2.1 Uniform Static Magnetic Field

The equation of motion is

$$\frac{d\mathbf{p}}{dt} = q\mathbf{v}\times\mathbf{B} \tag{7.8}$$

with $\mathbf{B}$ constant. Also, if W is the energy of the particle, W is constant, and thus both $v = |\mathbf{v}|$ and γ_v are constant. Thus $\frac{d\mathbf{p}}{dt} = \frac{d}{dt}(m\gamma_v\mathbf{v}) = m\gamma_v\frac{d\mathbf{v}}{dt}$ and the equation of motion becomes

$$\frac{d\mathbf{v}}{dt} = \mathbf{v}\times\boldsymbol{\omega}_B,$$

where

$$\boldsymbol{\omega}_B = \frac{q\mathbf{B}}{m\gamma_v} = \frac{q\mathbf{B}c^2}{W}.$$

If $\mathbf{B} = B\mathbf{k}$, and $\mathbf{v} = v_x\mathbf{i} + v_y\mathbf{j} + v_\parallel\mathbf{k}$, it follows that $v_\parallel$ is constant, and

$$\frac{dv_x}{dt} = v_y\omega_B,$$

$$\frac{dv_y}{dt} = -v_x\omega_B,$$

or

$$\frac{d}{dt}v_\perp = -iv_\perp\omega_B, \qquad (7.9)$$

where $v_\perp = v_x + iv_y$. Thus

$$v_\perp = e^{-i\omega_B t}v_{\perp 0} \qquad (7.10)$$

and the trajectory is given by

$$x + iy = (x_0 + iy_0) + \frac{iv_{\perp 0}}{\omega_B}(e^{-i\omega_B t} - 1), \quad z = z_0 + v_\parallel t. \qquad (7.11)$$

The motion is along a helix, winding around the lines of the B-field, with a radius $a = \frac{|v_{\perp 0}|}{\omega_B}$, the *radius of gyration*. Note that $|\mathbf{p}_\perp| = m\gamma_v|\mathbf{v}_\perp| = m\gamma_v a\omega_B = aqB$.

The curvature of the tracks of charged particles in a magnetic field can be used to determine the momentum of the particle. For this reason, the detectors used in high-energy particle physics experiments almost invariably incorporate magnets to bend the tracks of the particles produced in the processes under investigation. In addition to "bending magnets" that steer the beam of particles being accelerated in a synchrotron, there are other magnets, for example quadrupole or sextupole, used to focus the beam. And electron microscopes use both electrostatic and electromagnetic fields to act as lenses for the beam of electrons that is the analog of the light in an optical microscope. These applications of the Lorentz equation require careful design to shape the fields to fit their purpose.

7.2.2 Crossed E and B Fields

Consider now the motion of a charged particle in constant uniform perpendicular $\mathbf{E}$ and $\mathbf{B}$ fields. The equation of motion in the original reference frame K is

$$\frac{d\mathbf{p}}{dt} = q(\mathbf{E} + \mathbf{v} \times \mathbf{B}) \qquad (7.12)$$

or in a frame K' moving with velocity $\mathbf{u}$ with respect to K

$$\frac{d\mathbf{p}'}{dt'} = q(\mathbf{E}' + \mathbf{v}' \times \mathbf{B}'). \qquad (7.13)$$

If we now choose $\mathbf{u}$ perpendicular to both $\mathbf{E}$ and $\mathbf{B}$, for example, $\mathbf{E} = E\mathbf{j}$, $\mathbf{B} = B\mathbf{k}$, $\mathbf{u} = u\mathbf{i}$, we have

$$
\begin{aligned}
E'_x &= 0, \\
E'_y &= (E - Bc\beta_u)\gamma_u, \\
E'_z &= 0;
\end{aligned}
\tag{7.14}
$$

$$
\begin{aligned}
B'_x &= 0, \\
B'_y &= 0, \\
B'_z &= (B - E\beta_u/c)\gamma_u.
\end{aligned}
\tag{7.15}
$$

If $| E | < c | B |$ by choosing $\beta_u = \frac{E}{Bc}$ there results

$$
\mathbf{E}' = 0,
\tag{7.16}
$$

$$
\mathbf{B}' = \left(\frac{c^2 B^2 - E^2}{c^2 B^2} \right)^{\frac{1}{2}} \mathbf{B},
\tag{7.17}
$$

so that in K' we have just the case previously considered. Thus the motion in K has superposed on the helical motion around $\mathbf{B}$-lines a uniform drift with velocity $\mathbf{u}$ perpendicular to both $\mathbf{E}$ and $\mathbf{B}$.

If on the other hand $| E | > c | B |$ there is no frame in which $\mathbf{E}'$ vanishes, but there is a frame K' in which $\mathbf{B} = 0$, obtained by taking $\beta_u = cB/E$, and then in K' the motion is again easy to derive, etc.

Note that in the first case if the particle had a velocity $\mathbf{u} = \frac{E}{B}\mathbf{i}$ in K, its velocity in K' would be zero, and so would remain zero, and so back again in K the velocity would remain constant $= \mathbf{u}$. This is the basis of velocity selection devices. It was also famously used by J. J. Thomson (1856–1940) in his 1897 experiments to determine the ratio of charge to mass of the electron.

7.2.3   Nonuniform Static B-Field

Suppose now that $\mathbf{B} = B\mathbf{k}$ as previously considered, but that B is no longer constant. The interesting effect is when the nonuniformity of $\mathbf{B}$ is such that

$$
\nabla B = B\boldsymbol{\epsilon}
\tag{7.18}
$$

with $\boldsymbol{\epsilon}$ small, and orthogonal to $\mathbf{B}$; we may take it to be constant in the region of interest. We still have

$$
\frac{d\mathbf{v}}{dt} = \mathbf{v} \times \boldsymbol{\omega}
$$

with $\boldsymbol{\omega} = \frac{qB}{m\gamma}\mathbf{k}$ and constant v as before, but now B is no longer a constant. Let us describe the motion that would result for constant B by

$$
\mathbf{r}_0 = \mathbf{R} + \boldsymbol{\xi}_0
$$

in which $\mathbf{R}$ describes the uniformly drifting center around which the circular motion given by $\boldsymbol{\xi}_0$ takes place, that is, writing

$$
\mathbf{v} = \mathbf{v}_\perp + \mathbf{v}_\|
$$

we have
$$\dot{\mathbf{R}} = \mathbf{v}_{\parallel} = \text{constant}, \qquad \dot{\boldsymbol{\xi}}_0 = (\mathbf{v}_{\perp})_0 = -\boldsymbol{\omega}_0 \times \boldsymbol{\xi}_0.$$

Expanding $B(\mathbf{r})$ to first order about the point $\mathbf{R}$,
$$B(\mathbf{r}) = B(\mathbf{R}) + \boldsymbol{\xi} \cdot \nabla B(\mathbf{R}) + \cdots$$
$$\approx B_0(1 + \boldsymbol{\xi} \cdot \boldsymbol{\epsilon}),$$

so that to first order
$$\frac{d\mathbf{v}_{\perp}}{dt} = \mathbf{v}_{\perp} \times \boldsymbol{\omega}_0(1 + \boldsymbol{\xi} \cdot \boldsymbol{\epsilon}). \tag{7.19}$$

Now set $\boldsymbol{\xi} = \boldsymbol{\xi}_0 + \boldsymbol{\eta}$ where $\boldsymbol{\eta}$ is small. Again working to first order in small quantities,
$$\ddot{\boldsymbol{\eta}} = [\dot{\boldsymbol{\eta}} - (\boldsymbol{\omega}_0 \times \boldsymbol{\xi}_0)(\boldsymbol{\epsilon} \cdot \boldsymbol{\xi}_0)] \times \boldsymbol{\omega}_0. \tag{7.20}$$

The contribution of the first term is just like the original motion in the uniform field, and may be absorbed into that circular motion around the moving center $\mathbf{R}$. In the second and more interesting term, both factors $\boldsymbol{\xi}_0$ oscillate, and each has a vanishing time average. But their product does not have a vanishing time average, so that there is produced an additional drift velocity $\mathbf{v}_G$, the *gradient drift*, given by
$$\mathbf{v}_G = \langle \dot{\boldsymbol{\eta}} \rangle$$
$$= \langle (\boldsymbol{\omega}_0 \times \boldsymbol{\xi}_0)(\boldsymbol{\epsilon} \cdot \boldsymbol{\xi}_0) \rangle$$
$$= \frac{1}{2}a^2 \boldsymbol{\omega}_0 \times \boldsymbol{\epsilon}$$
$$= \frac{1}{2}a^2 \frac{\omega_B}{B^2}\mathbf{B} \times \nabla_{\perp} B. \tag{7.21}$$

7.2.4   Curved Magnetic Field Lines

In a region where the magnetic field lines are curved, so long as the curvature is gradual we may always suppose them to be along an arc of a (large) circle of radius R in the region of interest. By choosing cylindrical polar coordinates (ρ, ϕ, z) with axis through the centre of curvature of the field lines, and perpendicular to their plane, the field can be written as $\mathbf{B} = (0, B, 0)$, and then the equation of motion
$$\frac{d\mathbf{v}}{dt} = \frac{q}{m\gamma}\mathbf{v} \times \mathbf{B}$$

gives
$$\ddot{\rho} - \rho\dot{\phi}^2 = -\omega_B \dot{z},$$
$$\rho\ddot{\phi} + 2\dot{\rho}\dot{\phi} = 0,$$
$$\ddot{z} = \omega_B \dot{\rho}, \tag{7.22}$$

with $\omega_B = \frac{qB}{m\gamma}$ as usual. If in the absence of curvature to the field, the trajectory had been along a helix with $a \ll R$, these give
$$\dot{\phi} \approx \frac{v_{\parallel}}{R}, \qquad \rho \approx R,$$

and then it follows that

$$\ddot{z} \approx \frac{v_{\parallel}^2}{\omega_B R}.$$

That is, there is an additional drift velocity, the *curvature drift*

$$\mathbf{v}_C = \frac{v_{\parallel}^2}{\omega_B R} \frac{\mathbf{R} \times \mathbf{B}}{RB}. \tag{7.23}$$

Adding this to the previous result for the gradient drift gives the total drift velocity brought about by the nonuniformity in the field,

$$\mathbf{v}_D = \mathbf{v}_G + \mathbf{v}_C$$

$$= \frac{v_{\parallel}^2}{\omega_B R} \frac{\mathbf{R} \times \mathbf{B}}{RB} + \frac{1}{2} \frac{a}{B^2} \mathbf{B} \times \nabla_{\perp} B. \tag{7.24}$$

But $\nabla \times \mathbf{B} = 0$ in the absence of currents, which implies

$$\frac{\nabla_{\perp} B}{B} = -\frac{\mathbf{R}}{R^2},$$

and then

$$\mathbf{v}_D = \frac{1}{\omega_B R} (v_{\parallel}^2 + \frac{1}{2} v_{\perp}^2) \frac{\mathbf{R} \times \mathbf{B}}{RB}. \tag{7.25}$$

For particles in thermal equilibrium, this gives an average drift speed

$$v_D = \frac{2}{\omega R} \frac{k_B T}{m}, \tag{7.26}$$

or for nonrelativistic motion, working in SI units, $v_D = 1.7 \times 10^{-4} \frac{T}{RB}$. To get some idea of the importance of this effect in a realistic situation, consider the conditions in a thermonuclear fusion device such as ITER,[1] where $T \approx 10^8$ K to achieve fusion, $R \approx 13$ m as a typical dimension of the apparatus, and $B \approx 13$ T. Then one finds $v_D \approx 100$ m·s^{-1}, which is quite large, and contributes substantially to the design difficulties in containing the plasma.

7.3   Exercises for Chapter 7

1   Ionized atoms or molecules passing through a magnetic field are deflected, and the amount of deflection depends on the ratio of charge to mass of the ions. Explain how this is used in a mass spectrometer. If the instrument is to be used to determine the ratio of C^{13} to C^{12} in a sample of carbon dioxide, suggest appropriate design parameters that might be used if there is available an ionization chamber that provided a beam of carbon dioxide ions; this chamber is held at a high voltage, typically a few kilovolts. Also available is a magnet which provides a field of the order 1 T. Pay attention to the size of the components,

---

[1]ITER is an experimental thermonuclear reactor being constructed by an international collaboration at Cadarache in southern France.

and to the detection and instrumetation needed. Why do you need a vaccuum in the apparatus?

2   A "magnetic horn" is a device that is used to focus a beam of charged particles. It consists of two coaxial conductors that carry a current (typically several hundred kA) that generate a solenoidal magnetic field in the space between them.

(a) Show that the magnetic field intensity varies inversely with the distance from the horn axis.

(b) Show that this device is capable of focusing a beam of charged particles produced close to its axis by reducing the transverse (radial) component of their momenta.

[The horn used to focus pions that would then decay to give neutrinos for experiments at CERN was about 1 m long but with a much smaller radial size. The pions emerged from the target in which they were produced with a typical transverse momentum of 250 MeV/c.]

3   Explain how crossed electric and magnetic fields can be used to select charged particles with a particular velocity.

Chapter 8

Fields from Sources

In this and the following chapters, the main emphasis will be on radiation—how it is generated by moving charges and varying currents, and how it is scattered. In the book so far we have seen how the charge and current densities are affected by electromagnetic fields, and how the electromagnetic fields are affected by the charge and current densities. But we have yet to study how changing charge-current densities can generate electromagnetic *waves*, nor how such waves can be scattered, and so on.

8.1   Introducing the Green's Function

Let us go back to Maxwell's equations, or rather to the the equation relating the potentials to the sources:

$$\Box A^\mu - \partial^\mu(\partial_\nu A^\nu) = \mu_0 j^\mu,$$

or in Lorenz gauge simply

$$\Box A^\mu = \mu_0 j^\mu. \tag{8.1}$$

We want to solve this equation for suitable choices of the source term on the right-hand side. Note that this is an *inhomogeneous* differential equation, and as is true for such equations, we can add any solution to the corresponding *homogeneous* equation

$$\Box A^\mu = 0$$

and still have a solution. This is usually described as the addition of a *complementary function* to a *particular integral*. The choice of the complementary function to be used depends on the boundary conditions, or other requirements imposed by the physics of the situation being modeled.

For the time being, suppose that there are no boundary surfaces; the reason for this is that the boundary conditions which they would introduce would in general spoil the *linearity* of the problem that we now wish to exploit. Because of this linearity, we may make a Fourier analysis of the time dependence, and treat each component (i.e., each frequency) independently. So we write

$$j^\mu(\mathbf{x}, t) = \frac{1}{2\pi} \int_{-\infty}^{\infty} j^\mu(\mathbf{x}, \omega) e^{-i\omega t} \, d\omega, \tag{8.2}$$

$$A^\mu(\mathbf{x}, t) = \frac{1}{2\pi} \int_{-\infty}^{\infty} A^\mu(\mathbf{x}, \omega) e^{-i\omega t} \, d\omega, \tag{8.3}$$

and find

$$(\nabla^2 + k^2) A^\mu(\mathbf{x}, \omega) = -\mu_0 j^\mu(\mathbf{x}, \omega), \tag{8.4}$$

the inhomogeneous Helmholtz equation, where $k^2 c^2 = \omega^2$.

Again we may exploit the linearity by recognizing that a linear superposition of sources leads to a corresponding linear superposition of potentials; and so we are led to seek a solution for a special source, from which the general source can be built up by linear superposition:

$$(\nabla^2 + k^2) G_k(\mathbf{x}, \mathbf{x}') = -4\pi \delta^{(3)}(\mathbf{x} - \mathbf{x}'). \tag{8.5}$$

On the right-hand side, the source term is a delta function, representing a source concentrated at one point, $\mathbf{x} = \mathbf{x}'$. Before going further, we recall some of the basic properties of the delta function.

8.2 The Delta Function

The delta function is a continuum generalization of the Kronecker delta. Recall that this is defined by

$$\delta_{nn'} = \begin{cases} 1, & \text{if } n = n', \\ 0, & \text{if } n \neq n', \end{cases}$$

which implies

$$\sum_n \delta_{nn'} f_n = f_{n'}.$$

We seek a generalization from the discrete n to a continuous variable x:

$$f_n \to f(x),$$

for which

$$\int dx \, \delta(x - x') \, f(x) = f(x'). \tag{8.6}$$

This leads us to define $\delta(x)$ to satisfy

$$\delta(x - x') = 0, \qquad x \neq x',$$
$$\delta(x - x') = \delta(x' - x); \tag{8.7}$$

and for all suitably smooth functions $f(x)$,

$$\int_{-\infty}^{\infty} dx \, \delta(x - x') \, f(x) = f(x'). \tag{8.8}$$

It follows directly that

$$\int dx\, \delta(x - x') = 1$$

so long as the integration range covers the point $x = x'$. Also

$$\delta(cx) = \frac{1}{|c|}\delta(x).$$

We will also need

$$\delta(x^2 - a^2) = \frac{1}{2|a|}[\delta(x - a) + \delta(x + a)],$$

which is a special case of

$$\delta\big(g(x)\big) = \sum_i \frac{\delta(x - x_i)}{|g'(x_i)|},$$

where x_i are the roots of $g(x) = 0$.

A useful and straightforward generalization is to $\delta^{(3)}(\mathbf{x} - \mathbf{x}')$, which satisfies

$$\delta^{(3)}(\mathbf{x} - \mathbf{x}') = 0, \qquad \mathbf{x} \neq \mathbf{x}',$$
$$\delta^{(3)}(\mathbf{x} - \mathbf{x}') = \delta^{(3)}(\mathbf{x}' - \mathbf{x}),$$
$$\int d^3x\, \delta^{(3)}(\mathbf{x} - \mathbf{x}')\, f(\mathbf{x}) = f(\mathbf{x}'), \tag{8.9}$$

for any suitably smooth function f as long as the region of integration includes the point $\mathbf{x} = \mathbf{x}'$. With *Cartesian* coordinates $\mathbf{x} = (x, y, z)$, we have

$$\delta^{(3)}(\mathbf{x} - \mathbf{x}') = \delta(x - x')\delta(y - y')\delta(z - z'),$$

but for example in spherical polar coordinates $\mathbf{r} = (r, \theta, \phi)$,

$$\delta^{(3)}(\mathbf{r} - \mathbf{r}') \neq \delta(r - r')\delta(\theta - \theta')\delta(\phi - \phi').$$

In fact

$$\delta^{(3)}(\mathbf{r} - \mathbf{r}') = \frac{1}{r^2 \sin\theta}\delta(r - r')\delta(\theta - \theta')\delta(\phi - \phi'). \tag{8.10}$$

You may recognize the denominator as the *Jacobian* of the transformation from Cartesian to spherical polar coordinates, which enters into the equation

$$dx\, dy\, dz = r^2 \sin\theta\, dr\, d\theta\, d\phi.$$

8.3 The Green's Function

Let us return to the inhomogeneous Helmholtz equation with the delta function source:

$$(\nabla^2 + k^2)G_k(\mathbf{x}, \mathbf{x}') = -4\pi\delta^{(3)}(\mathbf{x} - \mathbf{x}').$$

A solution to this equation is called a *Green's function* for the differential operator $\nabla^2 + k^2$. Since we have excluded any boundary surfaces,[1] the solution will:

- depend only on $\mathbf{r} \equiv \mathbf{x} - \mathbf{x}'$;

- be spherically symmetrical, that is, will depend only on $r = |\mathbf{r}|$.

Changing variables to $\mathbf{r}$ expressed in spherical polars, we have

$$\frac{1}{r}\frac{d^2}{dr^2}(rG_k) + k^2 G_k = -4\pi\delta^{(3)}(\mathbf{r}). \qquad (8.11)$$

[Note $\delta^{(3)}(\mathbf{r}) \neq \delta(r)$!] Everywhere except at $r = 0$ we have

$$\frac{1}{r}\frac{d^2}{dr^2}(rG_k) + k^2 G_k = 0,$$

from which it follows that

$$rG_k = Ae^{ikr} + Be^{-ikr}, \qquad r \neq 0.$$

Also, for $r \to 0$,

$$\frac{1}{r}\frac{d^2}{dr^2}(rG_k) = -4\pi\delta^{(3)}(\mathbf{r}).$$

Now this is just what we would get from the Poisson equation

$$\nabla^2\Phi = -\frac{\rho}{\epsilon_0},$$

familiar from electrostatics, if the charge density is $\rho = 4\pi\epsilon_0\delta^{(3)}(\mathbf{r})$, corresponding to a point charge at the origin of magnitude $q = 4\pi\epsilon_0$. Since we know the solution to Poisson's equation is then

$$\Phi = \frac{q}{4\pi\epsilon_0}\frac{1}{r} = \frac{1}{r},$$

it follows that

$$G_k \to \frac{1}{r} \qquad \text{as} \quad kr \to 0. \qquad (8.12)$$

Thus

$$G_k = AG_k^{(+)} + BG_k^{(-)}, \qquad A + B = 1,$$

where

$$G^{(\pm)} = \frac{e^{\pm ikr}}{r}. \qquad (8.13)$$

---

[1] Although we have chosen not to pursue their inclusion, when boundary surfaces are present, the appropriate Green's function may be derived when the conditions to be satisfied at the boundaries are specified. These will often be of the *Dirichlet* kind, when the value of the Green's function at the boundary is specified, or of the *Neumann* kind, when it is the normal derivative of the Green's function that is specified.

Recall that we obtained the Helmholtz equation we are trying to solve as the equation associated with the frequency $\omega = ck$ of the original time-dependent equation. Therefore it will be multiplied by a factor $\exp(-i\omega t)$. So we consider the functions

$$G_k^{(\pm)} e^{-i\omega t} = \frac{e^{-i(\omega t \mp kr)}}{r}, \tag{8.14}$$

which represent spherical waves going out from/going in to the origin $\mathbf{r} = 0$, that is, the point $\mathbf{x} = \mathbf{x}_0$. Physical considerations will in most cases dictate the choice of an *outgoing* wave, so that we will most often choose the Green's function

$$G_k^{(+)} = \frac{e^{ikr}}{r},$$

or

$$G_k^{(+)}(\mathbf{x}, \mathbf{x}') = \frac{\exp[ik|\mathbf{x} - \mathbf{x}'|]}{|\mathbf{x} - \mathbf{x}'|}. \tag{8.15}$$

From the definition of the delta function it follows that

$$\mu_0 j^\mu(\mathbf{x}, \omega) = \frac{\mu_0}{4\pi} \int j^\mu(\mathbf{x}', \omega) \, 4\pi \delta^{(3)}(\mathbf{x} - \mathbf{x}') \, d^3x'.$$

From the linearity of our problem, as suggested at the end of Section 8.1, the solution to

$$(\nabla^2 + k^2) A^\mu(\mathbf{x}, \omega) = -\mu_0 j^\mu(\mathbf{x}, \omega)$$

which satisfies outgoing boundary conditions is to be built up by linear superposition from the Green's function, giving

$$A^\mu(\mathbf{x}, \omega) = \frac{\mu_0}{4\pi} \int j^\mu(\mathbf{x}', \omega) \frac{e^{ik|\mathbf{x} - \mathbf{x}'|}}{|\mathbf{x} - \mathbf{x}'|} \, d^3x'. \tag{8.16}$$

We now go from the frequency domain expression $j^\mu(\mathbf{x}', \omega)$ to the corresponding time domain expression $j^\mu(\mathbf{x}', t)$, and so obtain for the source term in Equation 8.16

$$j^\mu(\mathbf{x}', \omega) = \int_{-\infty}^{\infty} j^\mu(\mathbf{x}', t') e^{i\omega t'} \, dt',$$

which on substitution into Equation 8.16 gives

$$
\begin{aligned}
A^\mu(\mathbf{x}, t) &= \frac{1}{2\pi} \int_{-\infty}^{\infty} A^\mu(\mathbf{x}, \omega) e^{-i\omega t} \, d\omega \\
&= \frac{1}{2\pi} \int \left[\frac{\mu_0}{4\pi} j^\mu(\mathbf{x}', \omega) \frac{e^{ik|\mathbf{x} - \mathbf{x}'|}}{|\mathbf{x} - \mathbf{x}'|} \, d^3x' \right] e^{-i\omega t} \, d\omega \\
&= \frac{1}{2\pi} \int \left[\frac{\mu_0}{4\pi} \left(\int_{-\infty}^{\infty} j^\mu(\mathbf{x}', t') e^{i\omega t'} \, dt' \right) \frac{e^{ik|\mathbf{x} - \mathbf{x}'|}}{|\mathbf{x} - \mathbf{x}'|} \, d^3x' \right] e^{-i\omega t} \, d\omega. \tag{8.17}
\end{aligned}
$$

The ω-integration involves the Fourier transform of the Green's function,

$$\frac{1}{2\pi} \int_{-\infty}^{\infty} d\omega \, \frac{e^{ikr}}{r} e^{-i\omega(t-t')} = \frac{1}{2\pi} \int_{-\infty}^{\infty} d\omega \, \frac{\exp[i(\frac{r}{c} - \tau)\omega]}{r},$$

where $\tau = t - t'$ is the *relative time*; and we now may use

$$\frac{1}{2\pi} \int_{-\infty}^{\infty} d\omega\, e^{-i\lambda\omega} = \delta(\lambda) \qquad (8.18)$$

to write

$$
\begin{aligned}
G^{(+)}(\mathbf{x}, t; \mathbf{x}', t') &\equiv \frac{1}{2\pi} \int_{-\infty}^{\infty} d\omega\, G_k^{(+)}(\mathbf{x} - \mathbf{x}') e^{-i\omega(t-t')} \\
&= \frac{1}{2\pi} \int_{-\infty}^{\infty} d\omega\, \frac{e^{ik|\mathbf{x} - \mathbf{x}'|}}{|\mathbf{x} - \mathbf{x}'|} e^{-i\omega(t-t')} \\
&= \frac{\delta\left(t - (t' + |\mathbf{x}' - \mathbf{x}|/c)\right)}{|\mathbf{x}' - \mathbf{x}|} \\
&= \frac{\delta(\tau - r/c)}{r}.
\end{aligned}
\qquad (8.19)
$$

So we have obtained

$$A^\mu(\mathbf{x}, t) = \frac{\mu_0}{4\pi} \int d^3x' \int dt'\, G^{(+)}(\mathbf{x}, t; \mathbf{x}', t')\, j^\mu(\mathbf{x}', t') \qquad (8.20)$$

as a particular solution to our original problem

$$\Box A^\mu = \mu_0 j^\mu.$$

The response at $(\mathbf{x}, t)$ from the source at $(\mathbf{x}', t')$ is propagated by the Green's function

$$G^{(+)} = \frac{1}{r} \delta\left(t - (t' + r/c)\right),$$

where $r \equiv |\mathbf{x} - \mathbf{x}'|$, and occurs at a time $t = t' + r/c$ *retarded* by r/c, which is the time taken to travel the distance r at the speed of light. For this reason, $G^{(+)}$ is called the *retarded* Green's function (which is closely related to the Feynman *propagator* of relativistic quantum field theory).

We can perform the t' integration, and obtain

$$A_r^\mu(\mathbf{x}, t) = \frac{\mu_0}{4\pi} \int d^3x'\, j^\mu(\mathbf{x}', t - r/c)\frac{1}{r}, \qquad (8.21)$$

where again $r = |\mathbf{x} - \mathbf{x}'|$. These potentials are called the *retarded potentials*. In a similar fashion one can arrive at the *advanced* potentials

$$A_a^\mu(\mathbf{x}, t) = \frac{\mu_0}{4\pi} \int d^3x'\, j^\mu(\mathbf{x}', t + r/c)\frac{1}{r}.$$

8.4   The Covariant Form for the Green's Function

In writing

$$A_r^\mu = \frac{\mu_0}{4\pi} \int d^3x'\, j^\mu(\mathbf{x}', t - r/c)\frac{1}{r}$$

we have rather lost sight of the covariance of the theory. As a first step to making this apparent, write

$$A_r^\mu = \frac{\mu_0}{4\pi} \int d^4x' \frac{1}{c} \delta\left(t' - t + \frac{r}{c}\right) \frac{1}{r} j^\mu(x').$$ (8.22)

Now use

$$\begin{aligned}
\delta\left[(x - x')^2\right] &= \delta\left[c^2(t - t')^2 - (\mathbf{x} - \mathbf{x}')^2\right] \\
&= \delta\left[c^2(t - t')^2 - r^2\right] \\
&= \frac{1}{2r}\left[\delta\left(c(t - t') - r\right) + \delta\left[c(t - t') + r\right]\right] \\
&= \frac{1}{2r}\frac{1}{c}\left[\delta(t - t' - r/c) + \delta(t - t' + r/c)\right].
\end{aligned}$$ (8.23)

If we define

$$\theta(z) = \begin{cases} +1 & \text{if } z > 0; \\ 0 & \text{if } z < 0, \end{cases}$$

we see that

$$\begin{aligned}
\frac{1}{4\pi rc}\delta(t - t' - r/c) &= \frac{1}{2\pi}\delta\left[(x - x')^2\right]\theta(x_0 - x_0') \\
&\equiv D_r(x - x').
\end{aligned}$$ (8.24)

Although the theta function singles out the *time* difference, D_r is still Lorentz invariant, since a Lorentz transformation can never take a positive value of $x_0 - x_0'$ to a negative value if $(x - x')^2 \geq 0$. The expression for D_r thus shows that it is the *covariant* form for the *retarded* Green's function, and we have

$$A^\mu(x)_r = \mu_0 \int d^4x' \, D_r(x - x') \, j^\mu(x').$$ (8.25)

In an analogous way one has

$$A^\mu(x)_a = \mu_0 \int d^4x' \, D_a(x - x') \, j^\mu(x'),$$

which uses the *advanced* Green's function D_a.

8.5   Exercises for Chapter 8

1   A point charge q oscillates along the z-axis:

$$z = a \cos \omega t, x = 0, y = 0.$$

What are the electric and magnetic fields at large distances $r \gg b$?

2   A thin spherical shell carries a charge Q distributed uniformly over its surface. Its radius is a function of time: $r(t) = a(2 + \cos \omega t)$. Using equations for the retarded potentials expressed in spherical polar coordinates, find the fields at $r = R > 3a$ as functions of time.

3   The potentials outside a localized time-dependent source $j^\alpha(\mathbf{x}, t)$ may be expressed as an integral over the source:

$$A^\alpha(\mathbf{x}, t) = \frac{\mu_0}{4\pi} \int d^3\mathbf{x}' \, j^\alpha(\mathbf{x}', t_{\text{ret}}) \frac{1}{|\mathbf{x} - \mathbf{x}'|}.$$

(a)   Define t_{ret} and in a sentence or two indicate its physical significance.

A capacitor is made from two small spheres with centers at $+\mathbf{a}$ and $-\mathbf{a}$. Charges $+Q_0, -Q_0$ are placed on them. At time $t = 0$ a thin straight resistive wire is connected between them, so that they discharge. For $t > 0$ the charge is then

$$Q(t) = Q_0 e^{-t/\tau},$$

with a current

$$I(t) = \frac{Q_0}{\tau} e^{-t/\tau}$$

flowing from $+\mathbf{a}$ to $-\mathbf{a}$.

(b)   Show that for $ct > r \equiv |\mathbf{x}| \gg |\mathbf{a}|$, the scalar potential may be well approximated by

$$A^0(\mathbf{x}, t) = -\frac{\mu_0}{4\pi} cQ(t) \, 2\mathbf{a} \cdot \nabla \left(\frac{e^{r/c\tau}}{r} \right),$$

and $A^0(\mathbf{x}, t) = 0$ for $ct < r$.

The current density may likewise be approximated by

$$\mathbf{j}(\mathbf{x}, t) = -I(t) \, 2\mathbf{a} \delta^{(3)}(\mathbf{x}).$$

(c)   Determine the vector potential $\mathbf{A}(\mathbf{x}, t)$.
(d)   Verify that the potentials are in a Lorenz gauge.
(e)   Use

$$\mathbf{B} = \nabla \times \mathbf{A}, \quad \dot{\mathbf{E}} = c^2 \nabla \times \mathbf{B}, \quad \text{and} \quad \mathbf{S} = \frac{1}{\mu_0} \mathbf{E} \times \mathbf{B}$$

to show that for $r \gg c\tau \gg |\mathbf{a}|$,

$$\mathbf{S} = \frac{\mu_0}{4\pi^2} \frac{a^2 Q_0^2}{c\tau^4 r^2} \exp[-2(t - r/c)/\tau] \sin^2 \theta \, \mathbf{n}$$

for $ct > r$, and $\mathbf{S} = 0$ for $ct < r$. As usual $\mathbf{x} = n\mathbf{r}$ and θ is the angle between $\mathbf{n}$ and $\mathbf{a}$. [You may find it helpful to use $\nabla f(r) = \mathbf{n} f'(r)$. Also $[(\mathbf{a} \times \mathbf{n}) \times \mathbf{n}] \times (\mathbf{a} \times \mathbf{n}) = (\mathbf{a} \times \mathbf{n})^2 \mathbf{n}$.] [Reminder: $\mathbf{S}$ is the Poynting vector, the vector giving the energy flux carried by the electromagnetic field.]

(f)   By integrating $\mathbf{S}$ over a large sphere to determine the total flux of energy out from that sphere, and then integrating the result with respect to time, show that a fraction

$$\frac{8}{3} \left(\frac{a}{c\tau} \right)^3$$

of the energy initially stored in the capacitor is radiated away as the capacitor discharges.

Chapter 9

Radiation

We are now in a position to exploit the formalism we have developed to obtain results for the radiation from time-dependent charge and current distributions. But let us first confirm that the formalism reproduces some basic results for static sources.

If the source charge-current density j^μ is independent of time, i.e., it is static, then the resulting fields will also be time independent:

$$A^\mu(\mathbf{x}, t) = \frac{\mu_0}{4\pi} \int d^3x' \, j^\mu(\mathbf{x}') \frac{1}{r} = A^\mu(\mathbf{x}),$$

and

$$
\begin{aligned}
\mathbf{E}(\mathbf{x}) &= -\dot{\mathbf{A}} - \boldsymbol{\nabla}\Phi \\
&= -\boldsymbol{\nabla}(cA^0) \\
&= -\frac{1}{4\pi\epsilon_0} \boldsymbol{\nabla} \int d^3x' \, \rho(\mathbf{x}') \frac{1}{|\mathbf{x} - \mathbf{x}'|} \\
&= \frac{1}{4\pi\epsilon_0} \int \rho(\mathbf{x}') \frac{(\mathbf{x} - \mathbf{x}')}{|\mathbf{x} - \mathbf{x}'|^3} \, d^3x',
\end{aligned}
\tag{9.1}
$$

which is after all just Coulomb's law; and

$$
\begin{aligned}
\mathbf{B}(\mathbf{x}) &= \boldsymbol{\nabla} \times \mathbf{A} \\
&= \frac{\mu_0}{4\pi} \boldsymbol{\nabla} \times \int d^3x' \, \mathbf{j}(\mathbf{x}') \frac{1}{|\mathbf{x} - \mathbf{x}'|} \\
&= \frac{\mu_0}{4\pi} \int \mathbf{j}(\mathbf{x}') \times \frac{(\mathbf{x} - \mathbf{x}')}{|\mathbf{x} - \mathbf{x}'|^3} \, d^3x',
\end{aligned}
\tag{9.2}
$$

which is just the Biot-Savart law.

9.1   Potentials from a Moving Charged Particle

Suppose that the source of the fields is a point particle carrying a charge q which follows the trajectory $\mathbf{r} = \mathbf{r}(t)$. Then the source densities are

$$j^0 = cq\, \delta^{(3)}[\mathbf{x} - \mathbf{r}(t)],$$
$$\mathbf{j} = q\mathbf{u}(t)\, \delta^{(3)}[\mathbf{x} - \mathbf{r}(t)], \tag{9.3}$$

where $\mathbf{u} = \dot{\mathbf{r}} = \frac{d\mathbf{r}}{dt}$. If we recall that the particle's four-velocity U^μ is given by

$$U^\mu = (U^0 = \gamma_u c, \ \mathbf{U} = \gamma_u \mathbf{u}) = \gamma_u(c, \ \mathbf{u}),$$

these may be combined to give

$$j^\mu(\mathbf{x}, t) = q \frac{1}{\gamma_u} U^\mu \delta^{(3)}[\mathbf{x} - \mathbf{r}(t)]. \tag{9.4}$$

To make this manifestly covariant, we introduce the *proper time* τ along the particle's trajectory, that is, the time that would be measured out by a clock moving with the particle. As the particle moves, its position (in the original frame, K say) is given by $\mathbf{r}(t)$, but also the time interval $d\tau$ is related to dx^0 by

$$\frac{d\tau}{dx^0} = \frac{1}{c\gamma_u},$$

so that we may write

$$j^\mu(x) = cq \int d\tau\, U^\mu(\tau)\delta^{(4)}[x - r(\tau)], \tag{9.5}$$

where x is the four-vector with components x^α, and

$$r^\alpha(\tau) = [ct, \mathbf{r}(t)], \quad U^\alpha = [\gamma_u c, \gamma_u \mathbf{u}].$$

This follows from

$$c \int d\tau\, U^\mu(\tau)\delta[x^0 - r^0] = c \int \frac{d\tau}{dx^0}\, dx^0 U^\mu \delta(x^0 - ct) = \frac{1}{\gamma_u} U^\mu. \tag{9.6}$$

For the potentials we then have (applying outgoing boundary conditions):[1]

$$A^\mu(x) = \mu_0 \int d^4x'\, D_r(x - x')\, cq \int d\tau\, U^\mu(\tau)\delta^{(4)}[x' - r(\tau)]$$

$$= c\mu_0 q \int d\tau\, U^\mu(\tau) D_r[x - r(\tau)]$$

$$= \frac{c\mu_0 q}{2\pi} \int d\tau\, U^\mu(\tau)\delta\left\{ [x - r(\tau)]^2 \right\} \theta[x^0 - r^0(\tau)]. \tag{9.7}$$

The delta function then ensures that the field at x is determined by what happens on the trajectory of the particle only where $r(\tau)$ is lightlike separated from

---

[1] See Eqs.(8.24) and (8.25).

x. This means that the point $r(\tau)$ lies on the light cone with its vertex at x. There are *two* such points, one in the past of x, one to the future: the theta function then picks out the *unique* point at which the trajectory crosses the *past* light cone with vertex at x. When we come to evaluate $\int d\tau\, U(\tau)\delta[f(\tau)]$ there will in general be a sum over contributions from each of the roots of $f = 0$, but in the present case there is thus only the one contributing root, and we have

$$\int d\tau\, U(\tau)\delta[f(\tau)] = \int \frac{d\tau}{df} U(\tau)\delta(f)\, df = \left[\frac{U(\tau)}{|df/d\tau|}\right]_{f=0}.$$

Since in the present case $f = [x - r(\tau)]^2$, we have

$$\left|\frac{df}{d\tau}\right| = \left|-2[x - r(\tau)]_\beta \frac{dr^\beta}{d\tau}\right|$$

$$= \left|2[x - r(\tau)]_\beta U^\beta(\tau)\right|,$$

and thus

$$A^\mu(x) = \frac{c\mu_0 q}{4\pi}\left[\frac{U^\mu(\tau)}{U(\tau)\cdot[x - r(\tau)]}\right]_{\tau=\tau_0}, \tag{9.8}$$

where τ_0 is defined by

$$[x - r(\tau_0)]^2 = 0 \quad \text{and} \quad x^0 > r^0(\tau_0). \tag{9.9}$$

This is then manifestly covariant.

9.2 The Liénard-Wiechert Potentials

We can render this last expression more useful as follows. The light-cone condition means

$$[x - r(\tau_0)]^2 = [x^0 - r^0(\tau_0)]^2 - [\mathbf{x} - \mathbf{r}(\tau_0)]^2 = 0,$$

or in terms of $R \equiv |\mathbf{x} - \mathbf{r}(\tau_0)|$,

$$x^0 - r^0(\tau_0) = R, \tag{9.10}$$

where we have used the condition $x^0 > r^0(\tau_0)$ to fix the sign. Note that this is a complicated equation for τ_0 which determines the (unique) point to to the past of the field point x on the trajectory of the particle from which an influence propagating at the speed of light reaches the position $\mathbf{x}$ at the time $ct = x^0$. The corresponding time r^0/c on the trajectory is called the *retarded time*, t_{ret}. Thus

$$c(t - t_{\text{ret}}) = R. \tag{9.11}$$

We also have

$$U\cdot[x - r(\tau_0)] = U^0[x^0 - r^0(\tau_0)] - \mathbf{U}\cdot[\mathbf{x} - \mathbf{r}(\tau_0)]$$

$$= U^0 R - \mathbf{U}\cdot[\mathbf{x} - \mathbf{r}(\tau_0)]$$

$$= \gamma_u cR(1 - \boldsymbol{\beta}\cdot\mathbf{n}), \tag{9.12}$$

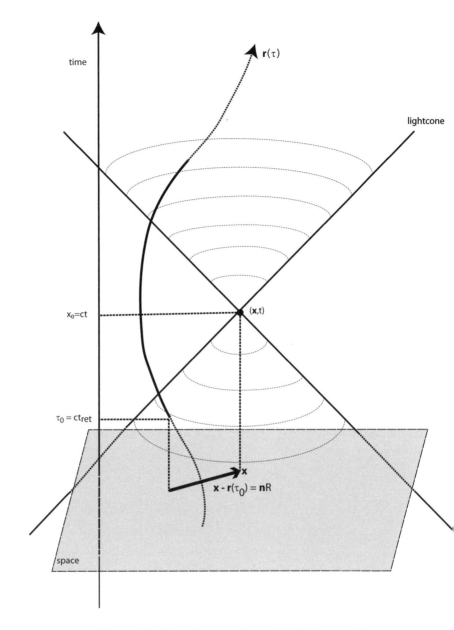

Figure 9.1. The trajectory of a particle, its intersection
with the past lightcone, and the definition of R and $\mathbf{r}(\tau_0)$.

where we have as usual written $\mathbf{u} = c\boldsymbol{\beta}$, and have defined the unit vector $\mathbf{n}$ that points from the point on the trajectory $\mathbf{r}(\tau_0)$ to the field point $\mathbf{x}$ by (Fig. 9.1)

$$\mathbf{x} - \mathbf{r}(\tau_0) = \mathbf{n}R. \tag{9.13}$$

This results in

$$A^\mu = \frac{c\mu_0 q}{4\pi} \left\{ \frac{[1, \boldsymbol{\beta}]}{R(1 - \boldsymbol{\beta} \cdot \mathbf{n})} \right\}_{\tau=\tau_0}. \tag{9.14}$$

In terms of the scalar Φ and vector $\mathbf{A}$, this is usually written as

$$\Phi = \frac{q}{4\pi\epsilon_0} \left[\frac{1}{R} \frac{1}{1 - \boldsymbol{\beta} \cdot \mathbf{n}} \right]_{\text{ret}},$$

$$\mathbf{A} = \frac{\mu_0 q}{4\pi} \left[\frac{\mathbf{u}}{R} \frac{1}{1 - \boldsymbol{\beta} \cdot \mathbf{n}} \right]_{\text{ret}}, \tag{9.15}$$

where the subscript "ret" indicates that the time variable to be used is the retarded time $t_{\text{ret}} = t - R/c$. These potentials are known as the *Liénard-Wiechert potentials*.

The Liénard-Wiechert potentials (or a direct calculation) may then be used to obtain

$$\mathbf{E} = \frac{q}{4\pi\epsilon_0} \left[\frac{\mathbf{n} - \boldsymbol{\beta}}{\gamma_u^2 R^2 (1 - \boldsymbol{\beta} \cdot \mathbf{n})^3} \right]_{\text{ret}} + \frac{q}{4\pi\epsilon_0} \frac{1}{c} \left[\frac{\mathbf{n} \times [(\mathbf{n} - \boldsymbol{\beta}) \times \dot{\boldsymbol{\beta}}]}{(1 - \boldsymbol{\beta} \cdot \mathbf{n})^3 R} \right]_{\text{ret}},$$

$$\mathbf{B} = \frac{1}{c} [\mathbf{n} \times \mathbf{E}]_{\text{ret}}, \tag{9.16}$$

in which

$$c\dot{\boldsymbol{\beta}} = \frac{d}{dt}\mathbf{u}$$

is the acceleration of the particle. The contribution proportional to this is called the *acceleration field*, the other term is called the *velocity field*. Note also that the velocity field falls away with increasing separation R like R^{-2}, while the acceleration field falls away more slowly, like R^{-1}; for this reason the acceleration field is dominant in what is called the far-field region.

9.2.1   Fields from an Unaccelerated Particle

This is the case already considered in Chapter 7, but we can now see how it follows as a special case of the approach in this section. The particle trajectory is

$$\mathbf{r} = c\boldsymbol{\beta}t,$$

with $\boldsymbol{\beta}$ constant, which we may take to be along the x-axis. Evidently there is no acceleration field. We want to find the fields at a point P that is distance b from the x-axis, and without loss of generality we take P on the y-axis, $P = b\mathbf{j}$. If the particle passes through the origin at time $t = 0$, its position at time t is at the point $Q = \beta ct\mathbf{i}$. At the retarded time (now written t_r) it was at $Q_r = \beta ct_r\mathbf{i}$. The time $t_r < t$ is determined by the condition that light at speed c could reach

P in the same time that the particle moving at speed βc moves from Q_r to Q. Since the distance R from Q_r to P is $\sqrt{b^2 + \beta^2 c^2 t_r^2}$, we have the condition

$$R^2 = c^2(t - t_r)^2 = b^2 + \beta^2 c^2 t_r^2. \tag{9.17}$$

The vector from Q_r to P is $R\mathbf{n}$, which defines the unit vector $\mathbf{n}$. To determine the fields we need to evaluate $R(1 - \boldsymbol{\beta} \cdot \mathbf{n}) = R + \beta^2 c t_r$, for which purpose we will use a geometric argument (Fig. 9.2). We define the point S on the line joining P to Q_r that is distance $R(1 - \boldsymbol{\beta} \cdot \mathbf{n})$ from P, and hence at distance $-\boldsymbol{\beta} \cdot \mathbf{n} R = \beta^2 c t_r$ from Q_r:

$$S = \beta c t_r \mathbf{i} + R(\boldsymbol{\beta} \cdot \mathbf{n}) \mathbf{n}. \tag{9.18}$$

Therefore

$$
\begin{aligned}
(SQ)^2 &= [\beta c(t_r - t)\mathbf{i} + R(\boldsymbol{\beta} \cdot \mathbf{n})\,\mathbf{n}]^2 \\
&= \beta^2 c^2 (t_r - t)^2 + R^2 (\boldsymbol{\beta} \cdot \mathbf{n})^2 + 2cR(t_r - t)(\boldsymbol{\beta} \cdot \mathbf{n})^2 \\
&= \beta^2 c^2 (t_r - t)^2 - R^2 (\boldsymbol{\beta} \cdot \mathbf{n})^2,
\end{aligned} \tag{9.19}
$$

where we have at the last step used $c(t - t_r) = R$. And since we have $(SQ_r)^2 = R^2(\boldsymbol{\beta} \cdot \mathbf{n})^2$,

$$(SQ)^2 + (SQ_r)^2 = \beta^2 c^2 (t - t_r)^2 = (QQ_r)^2. \tag{9.20}$$

From this, Pythagoras's theorem implies that QQ_r is the hypotenuse of a right-angled triangle, with the right angle at S.

Noting that $(PS)^2 = [R(1 - \boldsymbol{\beta} \cdot \mathbf{n})]^2$, and that PS is the hypotenuse of another right-angled triangle PSQ, we have $(PS)^2 = (PQ)^2 + (SQ)^2$. The triangles QSQ_r and OPQ_s are similar so that

$$(SQ) = (QQ_r)b/R = bc\beta(t - t_r)/R = b\beta.$$

It follows that

$$
\begin{aligned}
[R(1 - \boldsymbol{\beta} \cdot \mathbf{n})]^2 &= (b^2 + \beta^2 c^2 t^2) - \beta^2 b^2 \\
&= \frac{1}{\gamma^2}[b^2 + \beta^2 \gamma^2 c^2 t^2].
\end{aligned} \tag{9.21}
$$

We are now ready to put this into the expressions for the fields:

$$
\begin{aligned}
\mathbf{E} &= \frac{q}{4\pi\epsilon_0}\left[\frac{\mathbf{n} - \boldsymbol{\beta}}{\gamma^2 R^2 (1 - \boldsymbol{\beta} \cdot \mathbf{n})^3}\right] \\
&= \frac{q}{4\pi\epsilon_0}[b^2 + \beta^2 \gamma^2 c^2 t^2]^{-3/2}\gamma R(\mathbf{n} - \boldsymbol{\beta}) \\
&= \frac{q}{4\pi\epsilon_0}[b^2 + \beta^2 \gamma^2 c^2 t^2]^{-3/2}\gamma(-\beta c t_r \mathbf{i} + b\mathbf{j} - R\beta\mathbf{i}) \\
&= \frac{q}{4\pi\epsilon_0}[b^2 + \beta^2 \gamma^2 c^2 t^2]^{-3/2}\gamma\left(((-\beta c t_r - \beta c(t - t_r))\,\mathbf{i} + b\mathbf{j}\right) \\
&= \frac{q}{4\pi\epsilon_0}[b^2 + \beta^2 \gamma^2 c^2 t^2]^{-3/2}\gamma(-\beta c t\mathbf{i} + b\mathbf{j});
\end{aligned} \tag{9.22}
$$

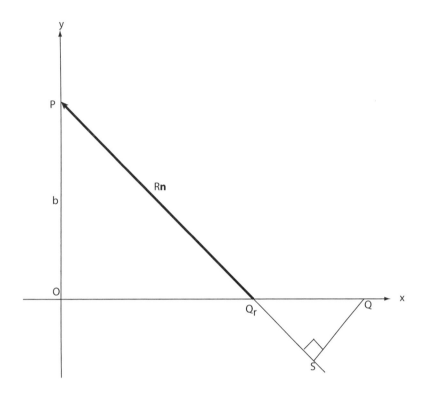

Figure 9.2. The geometric construction.

and

$$\mathbf{B} = \frac{1}{c}[\mathbf{n} \times \mathbf{E}]$$

$$= \frac{1}{c}\frac{q}{4\pi\epsilon_0}\frac{(-\beta ct_r\mathbf{i} \times b\mathbf{j})}{\gamma^2 R^3(1 - \boldsymbol{\beta}\cdot\mathbf{n})^3}$$

$$= -\frac{q}{4\pi\epsilon_0}\beta\gamma[b^2 + \beta^2\gamma^2 c^2 t^2]^{-\frac{3}{2}}b\mathbf{k}. \tag{9.23}$$

These do indeed agree with what was found in Chapter 7.

9.2.2 Fields from a Charged Oscillator

An important special case is that of the radiation from a charge executing simple harmonic motion—the charged *oscillator*. Let us write

$$\mathbf{r}(t) = \mathbf{a}\,e^{-i\omega t},$$

where we use the usual convention that the real part is to be taken, so that this really means $\mathbf{r} = \mathbf{a}\cos\omega t$. Then

$$\boldsymbol{\beta} = -\frac{\mathbf{a}}{c}i\omega\,e^{-i\omega t},$$

$$\dot{\boldsymbol{\beta}} = -\frac{\mathbf{a}}{c}\omega^2\,e^{-i\omega t}. \tag{9.24}$$

Let us also suppose $|\mathbf{a}|\omega \ll c$, so that the oscillator's motion is never relativistic. Then at all times $1 - \boldsymbol{\beta}\cdot\mathbf{n} \approx 1$. Also, if we consider the field far from the oscillator, where $|\mathbf{x}| \gg |\mathbf{a}|$, the distance R is always well approximated by $|\mathbf{x}|$. We then find

$$\mathbf{B} = \frac{1}{c}\mathbf{n} \times \mathbf{E},$$

$$\mathbf{E} \approx \frac{q}{4\pi\epsilon_0}\left[\frac{\mathbf{n}}{R^2} + \frac{1}{c}\mathbf{n} \times (\mathbf{n} \times \mathbf{a})\left(-\frac{\omega^2}{c}\right)\frac{e^{-i\omega t}}{R}\right]. \tag{9.25}$$

In these equations $\mathbf{x} = R\mathbf{n}$. The term proportional to $\frac{1}{R^2}$ is called the *near field*, and that proportional to $\frac{1}{R}$ is the *far* or *radiation field*. The *Poynting vector* is

$$\mathbf{S} = \frac{1}{\mu_0}\mathbf{E} \times \mathbf{B}$$

$$= \frac{1}{\mu_0}\left[\frac{q}{4\pi\epsilon_0}\left(-\frac{\omega^2}{c^2}\right)\frac{\cos\omega t}{R}\right]^2\frac{1}{c}[\mathbf{n} \times (\mathbf{n} \times \mathbf{a})] \times \{\mathbf{n} \times [\mathbf{n} \times (\mathbf{n} \times \mathbf{a})]\}$$

$$= \frac{\mu_0 q^2\omega^4\cos^2\omega t}{16\pi^2 cR^2}[a^2 - (\mathbf{n} \cdot \mathbf{a})^2]\mathbf{n}$$

$$= \frac{\mu_0 q^2}{16\pi^2 c}\frac{a^2}{R^2}\omega^4\cos^2\omega t\sin^2\theta\,\mathbf{n}, \tag{9.26}$$

where θ is the polar angle of the field point $\mathbf{x}$ with respect to a polar axis along the direction of the oscillation of the charge $\mathbf{a}$. (Fig. 9.3) A very similar result is obtained for the radiation from an oscillating *dipole*, and the $\sin^2\theta$ and ω^4 dependence are characteristic of dipole radiation.

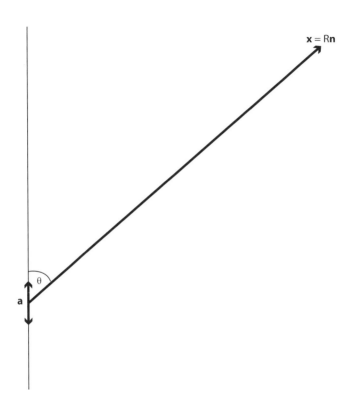

Figure 9.3. The charge oscillates along the polar axis.

9.3   The General Case

It is easy to see that for *general* accelerated motion of a charged particle, but considering only the radiation or far field,

$$\mathbf{S}_{\text{far}} = \frac{1}{\mu_0 c} \mathbf{E}_{\text{far}} \times [\mathbf{n} \times \mathbf{E}_{\text{far}}],$$

or, since $\mathbf{n} \cdot \mathbf{E}_{\text{far}} = 0$,

$$\mathbf{S}_{\text{far}} = \frac{1}{\mu_0 c} |\mathbf{E}_{\text{far}}|^2 \mathbf{n}. \tag{9.27}$$

The *power* radiated per unit solid angle is then

$$\frac{dP}{d\Omega} = \frac{1}{\mu_0 c} |R\mathbf{E}_{\text{far}}|^2. \tag{9.28}$$

If furthermore the motion of the particle is nonrelativistic, $\beta \ll 1$, we have

$$R\mathbf{E}_{\text{far}} = \frac{q}{4\pi\epsilon_0} \frac{1}{c} [\mathbf{n} \times (\mathbf{n} \times \dot{\boldsymbol{\beta}})],$$

so that

$$\begin{aligned}
\frac{dP}{d\Omega} &= \frac{1}{\mu_0 c} \left(\frac{q}{4\pi\epsilon_0} \right)^2 \frac{1}{c^2} [\mathbf{n} \times (\mathbf{n} \times \dot{\boldsymbol{\beta}})]^2 \\
&= \frac{q^2}{4\pi\epsilon_0} \frac{1}{4\pi c^3} |\dot{\mathbf{u}}|^2 \sin^2 \theta,
\end{aligned} \tag{9.29}$$

where θ is the angle between the direction $\mathbf{n}$ of the field point and the instantaneous acceleration $\dot{\mathbf{u}}$ of the particle (Fig. 9.4); and the radiation is polarized in the plane containing $\mathbf{n}$ and $\dot{\mathbf{u}}$. The total instantaneous power radiated is obtained by integration over angles, and using $\int d\Omega \sin^2 \theta = 8\pi/3$ one obtains the *Larmor formula*:

$$P = \frac{2}{3} \frac{q^2}{4\pi\epsilon_0} \frac{1}{c^3} |\dot{\mathbf{u}}|^2. \tag{9.30}$$

The Larmor formula has used $\beta \ll 1$, but if written in the form

$$P = \frac{2}{3} \frac{q^2}{4\pi\epsilon_0} \frac{1}{c^3} \frac{1}{m^2} \left(\frac{d\mathbf{p}}{dt} \cdot \frac{d\mathbf{p}}{dt} \right)$$

where $\mathbf{p} = m\mathbf{u}$ is the momentum of the charged particle, it generalizes to

$$\begin{aligned}
P &= -\frac{2}{3} \frac{q^2}{4\pi\epsilon_0} \frac{1}{c^3} \frac{1}{m^2} \left(\frac{dp_\mu}{d\tau} \frac{dp^\mu}{d\tau} \right) \\
&= \frac{2}{3} \frac{q^2}{4\pi\epsilon_0} \frac{1}{c} \gamma^6 [(\dot{\boldsymbol{\beta}})^2 - (\boldsymbol{\beta} \times \dot{\boldsymbol{\beta}})^2],
\end{aligned} \tag{9.31}$$

which is the *Liénard formula*.

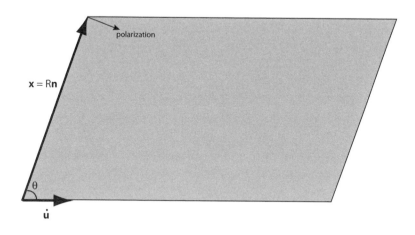

Figure 9.4. The vectors **n** and **u̇** define the plane of the polarization vector.

9.4   The Multipole Expansion

In electrostatics, the electrostatic potential $\Phi(\mathbf{x})$ at a point $\mathbf{x}$ outside a distribution ρ of charge is given by

$$\Phi(\mathbf{x}) = \frac{1}{4\pi\epsilon_0} \int \frac{\rho(\mathbf{x}')}{|\mathbf{x} - \mathbf{x}'|} \, d^3x'.$$

With $\mathbf{x}$ and $\mathbf{x}'$ expressed in polar coordinates, r, θ, ϕ and r', θ', ϕ', respectively, it is often useful to use the identity

$$\frac{1}{|\mathbf{x} - \mathbf{x}'|} = 4\pi \sum_{l=0}^{\infty} \sum_{m=-l}^{l} \frac{1}{2l+1} \frac{(r')^l}{r^{l+1}} Y_{lm}^*(\theta', \phi') Y_{lm}(\theta, \phi)$$

which uses the spherical harmonic functions Y_{lm}. This has the advantage of separating the $\mathbf{x}$ and $\mathbf{x}'$ variables enabling one to write

$$\Phi(\mathbf{x}) = \frac{1}{\epsilon_0} \sum_{l,m} \frac{1}{2l+1} Q_{lm} \frac{Y_{lm}(\theta, \phi)}{r^{l+1}},$$

where the coefficients Q_{lm} are given by

$$Q_{lm} = \int Y_{lm}^*(\theta', \phi')(r')^l \rho(\mathbf{x}') \, d^3x'.$$

These coefficients are *moments* of the charge distribution ρ, and the expansion is called the *multipole* expansion. The successive terms for $l = 0, l = 1, l = 2, \ldots$ are called the *monopole, dipole, quadrupole, $\ldots$* moments. We will exploit this expansion in what follows.

So let us now return to the general problem of the radiation from a localized system of charge and current densities $\rho(\mathbf{x}, t)$, $\mathbf{j}(\mathbf{x}, t)$, and recognize that we may make a Fourier time analysis to obtain a superposition of

$$\rho_\omega(\mathbf{x}, t) = \rho(\mathbf{x})e^{-i\omega t},$$

$$\mathbf{j}_\omega(\mathbf{x}, t) = \mathbf{j}(\mathbf{x})e^{-i\omega t}.$$

Then, working for convenience in a Lorenz gauge, one has for the vector potential $\mathbf{A}$ a corresponding superposition of frequencies:

$$\mathbf{A}_\omega(\mathbf{x}, t) = \frac{\mu_0}{4\pi} \int d^3x' \int dt' \frac{\mathbf{j}(\mathbf{x}', t')}{|\mathbf{x}' - \mathbf{x}|} \delta\left(t' + \frac{|\mathbf{x} - \mathbf{x}'|}{c} - t\right),$$

from which follows

$$\mathbf{A}_\omega(\mathbf{x}, t) = \mathbf{A}(\mathbf{x})e^{-i\omega t}$$

with

$$\mathbf{A}(\mathbf{x}) = \frac{\mu_0}{4\pi} \int d^3x' \, \mathbf{j}(\mathbf{x}') \frac{e^{ik|\mathbf{x}-\mathbf{x}'|}}{|\mathbf{x} - \mathbf{x}'|}, \tag{9.32}$$

where $k = \omega/c$. Outside the source, where $\mathbf{j} = 0$, we have $\frac{\partial \mathbf{E}}{\partial t} = c^2 \nabla \times \mathbf{B}$. Then from the harmonic time dependence $\frac{\partial \mathbf{E}}{\partial t} = -i\omega \mathbf{E} = -ikc\mathbf{E}$, we have

$$\mathbf{E} = \frac{ic}{k} \nabla \times \mathbf{B},$$

and as ever $\mathbf{B} = \nabla \times \mathbf{A}$.

Suppose that the sources are confined in a region of linear dimension $\approx d$, and that we are interested in the fields at distances $r \gg d$, that is, $|\mathbf{x}| \gg |\mathbf{x}'|$. The wavelength λ corresponding to the frequency ω is $\lambda = \frac{2\pi c}{\omega}$, and we will further suppose that the linear dimension of the source is small compared with the wavelength, $d \ll \lambda$. It is customary to define the *near zone* by $d \ll r \ll \lambda$, i.e., distances large compared with the size of the source, but small compared with the wavelength; the *intermediate zone* where $d \ll r \approx \lambda$; and the *far zone*, where $d \ll \lambda \ll r$, which is the zone of most interest. In the near zone, $kr \ll 1$, and also $e^{ik|\mathbf{x}-\mathbf{x}'|} \approx 1$, so that

$$\mathbf{A} \approx \frac{\mu_0}{4\pi} \int d^3x' \, \mathbf{j}(\mathbf{x}') \frac{1}{|\mathbf{x}' - \mathbf{x}|}.$$

The near field is *quasi-stationary*; it oscillates harmonically with a time dependence $e^{-i\omega t}$, but is otherwise just the same as for the stationary case.

In the far zone, many wavelengths away from the source which is itself small compared with a wavelength, we have $kr \gg 1 \gg kd$. We may use

$$\begin{aligned}
|\mathbf{x} - \mathbf{x}'| &= [(\mathbf{x} - \mathbf{x}')^2]^{1/2} \\
&= [r^2 - 2r\mathbf{n} \cdot \mathbf{x}' + \mathbf{x}'^2]^{1/2} \\
&= r - \mathbf{n} \cdot \mathbf{x}' + \cdots,
\end{aligned} \tag{9.33}$$

where $\mathbf{x} = \mathbf{n}r$ and the omitted terms are proportional to higher powers of $|\mathbf{x}'|/r \leq d/r$. Keeping only the lowest-order contributions, we have then[2]

$$\begin{aligned}
\mathbf{A} &\approx \frac{\mu_0}{4\pi} \int d^3x' \, \mathbf{j}(\mathbf{x}') \frac{\exp[ik(r - \mathbf{n} \cdot \mathbf{x}')]}{r} \\
&= \frac{\mu_0}{4\pi} \frac{e^{ikr}}{r} \int d^3x' \, \mathbf{j}(\mathbf{x}') \, e^{-ik\mathbf{n}\cdot\mathbf{x}'} \\
&= \frac{\mu_0}{4\pi} \frac{e^{ikr}}{r} \sum_{m=0}^{\infty} \frac{(-ik)^m}{m!} \int d^3x' \, \mathbf{j}(\mathbf{x}') \, (\mathbf{n} \cdot \mathbf{x}')^m.
\end{aligned} \tag{9.34}$$

Because the source is small, we may approximate further by neglecting all but

---

[2] Again an example of a multipole expansion.

the first contributing term in the power series, and in general will then obtain

$$\mathbf{A} \approx \frac{\mu_0}{4\pi} \frac{e^{ikr}}{r} \int d^3x' \, \mathbf{j}(\mathbf{x}')$$

$$= -\frac{\mu_0}{4\pi} \frac{e^{ikr}}{r} \int d^3x' \, \mathbf{x}'[\nabla' \cdot \mathbf{j}(\mathbf{x}')] \qquad \text{on integrating by parts}$$

$$= -\frac{\mu_0}{4\pi} \frac{e^{ikr}}{r} \int d^3x' \, \mathbf{x}'(i\omega)\rho(\mathbf{x}') \qquad \text{using the continuity equation}$$

$$= -\frac{\mu_0}{4\pi} \frac{e^{ikr}}{r} \, i\omega \, \mathbf{p}, \tag{9.35}$$

where $\mathbf{p}e^{-i\omega t} \equiv \int d^3x' \, \mathbf{x}' \, \rho(\mathbf{x}')e^{-i\omega t}$ is the *electric dipole moment* of the source. The dominant contribution to the fields in the far zone for a small source thus comes from the (oscillating) electric dipole moment. If the source were indeed just an oscillating dipole, the above expression for the vector potential would be exact.

It is interesting to see what we get by a similar calculation using the scalar potential. Since we have

$$\Phi(\mathbf{x}, t) = \frac{1}{4\pi\epsilon_0} \int d^3x' \int dt' \frac{\rho(\mathbf{x}')e^{-i\omega t'}}{|\mathbf{x}' - \mathbf{x}|} \delta\left(t' + \frac{|\mathbf{x}' - \mathbf{x}|}{c} - t\right),$$

an expansion in powers of $|\mathbf{x}'|/r$ is again possible. The zeroth-order term obtained by replacing $|\mathbf{x}' - \mathbf{x}|$ by r is

$$\Phi_{\text{monopole}} = \frac{1}{4\pi\epsilon_0} \int d^3x' \frac{\rho(\mathbf{x}')e^{-i\omega t'}}{r},$$

with $t' = t - r/c$. But this is just the *Coulomb potential* produced by a charge $Q = \int d^3x' \, \rho(\mathbf{x}', t')$ placed at the origin, and this total charge is *constant*. That is,

$$\Phi_{\text{monopole}} = \frac{Q}{4\pi\epsilon_0} \frac{1}{r}, \tag{9.36}$$

and there is no resultant radiation in zeroth order. There is no "electric monopole" radiation;[3] but the dipole and higher-moment terms enter in successive orders of approximation.

9.4.1   Electric Dipole Radiation

The electric dipole contribution is the lowest-order contribution in the systematic expansion multipole expansion of the radiation from the general source.

For the case of pure electric dipole radiation, we found

$$\mathbf{A} = -\frac{1}{4\pi\epsilon_0} \frac{e^{ikr}}{r} \frac{ik}{c} \mathbf{p},$$

---

[3]Nor indeed is there any "magnetic monopole" radiation—there are, so far as is now known, no magnetic monopoles, but even if they do exist, one may speculate that the total "magnetic charge" would be conserved in the same way as is true for the electric charge, and then an argument analogous to the above would show the absence of magnetic monopole radiation.

giving

$$\mathbf{B} = \nabla \times \mathbf{A}$$

$$= -\frac{1}{4\pi\epsilon_0}\frac{ik}{c}\left[\nabla\left(\frac{e^{ikr}}{r}\right)\right]\times\mathbf{p}$$

$$= \frac{1}{4\pi\epsilon_0}\frac{k^2}{c}\frac{e^{ikr}}{r}\left(1-\frac{1}{ikr}\right)\mathbf{n}\times\mathbf{p}. \tag{9.37}$$

And then from this we find

$$\mathbf{E} = \frac{ic}{k}\nabla\times\left[\frac{1}{4\pi\epsilon_0}\frac{k^2}{c}\frac{e^{ikr}}{r}\left(1-\frac{1}{ikr}\right)\mathbf{n}\times\mathbf{p}\right]$$

$$= \frac{ic}{k}\frac{1}{4\pi\epsilon_0}\frac{k^2}{c}\left\{\frac{e^{ikr}}{r}\left(1-\frac{1}{ikr}\right)\nabla\times(\mathbf{n}\times\mathbf{p})-(\mathbf{n}\times\mathbf{p})\times\mathbf{n}\frac{\partial}{\partial r}\left[\frac{e^{ikr}}{r}\left(1-\frac{1}{ikr}\right)\right]\right\}$$

$$= \frac{ik}{4\pi\epsilon_0}\frac{e^{ikr}}{r}\left\{-\frac{1}{r}[\mathbf{p}+(\mathbf{p}\cdot\mathbf{n})\mathbf{n}]\left(1-\frac{1}{ikr}\right)+[(\mathbf{n}\cdot\mathbf{p})\mathbf{n}-\mathbf{p}]\left(-\frac{2}{r}+\frac{2}{ikr^2}\right)\right\}$$

$$= \frac{k^2}{4\pi\epsilon_0}\frac{e^{ikr}}{r}(\mathbf{n}\times\mathbf{p})\times\mathbf{n}+\frac{1}{4\pi\epsilon_0}\frac{e^{ikr}}{r}\left(\frac{1}{r^2}-\frac{ik}{r}\right)[3\mathbf{n}(\mathbf{n}\cdot\mathbf{p})-\mathbf{p}]. \tag{9.38}$$

In the near zone $kr \ll 1$ these give

$$\mathbf{B} = \frac{1}{4\pi\epsilon_0}\frac{ik}{c}\mathbf{n}\times\mathbf{p}\frac{1}{r^2}, \tag{9.39}$$

$$\mathbf{E} = \frac{1}{4\pi\epsilon_0}[3\mathbf{n}(\mathbf{n}\cdot\mathbf{p})-\mathbf{p}]\frac{1}{r^3}, \tag{9.40}$$

which will be recognized as the expected (quasistatic—remember that the $e^{-i\omega t}$ time dependence has been omitted) fields from the electric dipole and the concomittant current.

In the far zone $kr \gg 1$ we have

$$\mathbf{B} = \frac{1}{4\pi\epsilon_0}\frac{k^2}{c}\frac{e^{ikr}}{r}\mathbf{n}\times\mathbf{p}, \tag{9.41}$$

$$\mathbf{E} = c\mathbf{B}\times\mathbf{n}. \tag{9.42}$$

As usual $\mathbf{E}$, $\mathbf{B}$, and $\mathbf{n}$ are mutually perpendicular. Note that the fields fall off like $\frac{1}{r}$. Putting back the (real) time dependence, we have for the Poynting vector[4]

$$\mathbf{S} = \frac{1}{\mu_0}[\Re(\mathbf{E}e^{-i\omega t})\times\Re(\mathbf{B}e^{-i\omega t})], \tag{9.43}$$

and for its *time average*, that is, the time average energy flux,

$$\langle\mathbf{S}\rangle = \frac{1}{2}\frac{1}{\mu_0}\Re[\mathbf{E}\times\mathbf{B}^*] = \frac{1}{2}\frac{1}{\mu_0 c}|\mathbf{E}|^2\mathbf{n}. \tag{9.44}$$

---

[4]We use the symbol $\Re$ to denote the real part of an expression.

And also, after time averaging,

$$\left\langle \frac{dP}{d\Omega} \right\rangle = r^2 \mathbf{n} \cdot \langle \mathbf{S} \rangle$$

$$= \frac{1}{2\mu_0 c} |r\mathbf{E}|^2$$

$$= \frac{1}{2\mu_0 c} \left(\frac{1}{4\pi\epsilon_0} k^2 \mathbf{n} \times \mathbf{p} \right)^2$$

$$= \frac{1}{2} \frac{\mu_0}{16\pi^2 c} |\mathbf{p}|^2 \omega^4 \sin^2 \theta, \tag{9.45}$$

which should be compared with the Larmor formula. Note again the characteristic $\sin^2 \theta$ dipole angular dependence and the ω^4 frequency behavior.

9.4.2   Magnetic Dipole and Higher-Order Terms

We may need to take into account further terms in the expansion beyond the electric dipole term, and will certainly need to do so if $\mathbf{p} = 0$. After the electric dipole contribution (which we may write as $\mathbf{A}_{(1)}$), the next term (including also for consistency the next term in the expansion of $|\mathbf{x} - \mathbf{x}'|^{-1}$) is

$$\mathbf{A}_{(2)}(\mathbf{x}) = \frac{\mu_0}{4\pi} \frac{e^{ikr}}{r} \left(\frac{1}{r} - ik \right) \int d^3x' \, \mathbf{j}(\mathbf{x}')(\mathbf{n} \cdot \mathbf{x}').$$

Let us write

$$\mathbf{j}(\mathbf{x}')(\mathbf{n} \cdot \mathbf{x}') = \frac{1}{2} \left[\mathbf{j}(\mathbf{x}')(\mathbf{n} \cdot \mathbf{x}') + (\mathbf{n} \cdot \mathbf{j}(\mathbf{x}')) \, \mathbf{x}' \right] + \frac{1}{2} \left[\mathbf{x}' \times \mathbf{j}(\mathbf{x}') \right] \times \mathbf{n}.$$

The reason for doing this becomes apparent when we recognize that the *second* of these terms gives rise to

$$\mathbf{A}_{\text{mag. dip.}} = ik \frac{\mu_0}{4\pi} \frac{e^{ikr}}{r} \left(1 - \frac{1}{ikr} \right) \mathbf{n} \times \mathbf{m}, \tag{9.46}$$

where

$$\mathbf{m} \equiv \frac{1}{2} \int d^3x' \left[\mathbf{x}' \times \mathbf{j}(\mathbf{x}') \right] \tag{9.47}$$

is the *magnetic dipole moment* of the source. The other contribution is then

$$\mathbf{A}_{\text{el.quad.}} = \frac{\mu_0}{4\pi} \frac{e^{ikr}}{r} \left(\frac{1}{r} - ik \right) \frac{1}{2} \int d^3x' \left[\mathbf{j}(\mathbf{x}')(\mathbf{n} \cdot \mathbf{x}') + (\mathbf{n} \cdot \mathbf{j})\mathbf{x}' \right]$$

$$= \frac{\mu_0}{4\pi} \frac{e^{ikr}}{r} \left(\frac{1}{r} - ik \right) \frac{-ikc}{2} \int d^3x' \, \mathbf{x}'(\mathbf{n} \cdot \mathbf{x}')\rho(\mathbf{x}') \tag{9.48}$$

(where again there has been an integration by parts, and use is made of the continuity equation to go from the first line to the next), which evidently is proportional to the second moment of the charge distribution, and so may be identified with the contribution of the *electric quadrupole*.

The fields corresponding to the previously obtained expression for the vector potential arising from the magnetic dipole may be obtained by substituting

$$\mathbf{E} \to c\mathbf{B}, \quad \mathbf{B} \to \frac{\mathbf{E}}{c}, \quad \mathbf{p} \to \frac{\mathbf{m}}{c}$$

in the expressions given above for the electric dipole fields. The angular and frequency dependence are thus unaltered, but the polarization is different. Whereas the electric dipole radiation was polarized with the electric field in the plane of $\mathbf{n}$ and $\mathbf{p}$, the magnetic dipole radiation has its electric field perpendicular to the plane defined by $\mathbf{n}$ and $\mathbf{m}$.

We will not derive the angular dependence, and so on, for the electric quadrupole term, nor the contributions from higher-order terms (electric sextupole/magnetic quadrupole, etc.), but the methodology is straightforward enough.

9.5   Motion in a Circle

Specializing now to the case of motion in a circle, where

$$\left|\frac{d\mathbf{p}}{d\tau}\right| = \left|\gamma\frac{d\mathbf{p}}{dt}\right| = \gamma\omega|\mathbf{p}|$$

if the energy loss per revolution is small

$$\frac{1}{c}\frac{dE}{d\tau} \ll \left|\frac{d\mathbf{p}}{d\tau}\right|,$$

we have

$$\begin{aligned}
P &= \frac{2}{3}\frac{q^2}{4\pi\epsilon_0}\frac{1}{c^3}\frac{1}{m^2}\gamma^2\omega^2|\mathbf{p}|^2 \\
&= \frac{2}{3}\frac{q^2}{4\pi\epsilon_0}\frac{1}{c}\frac{1}{m^2}\gamma^2\omega^2\gamma^2\beta^2 m^2 \\
&= \frac{2}{3}\frac{q^2}{4\pi\epsilon_0}c\beta^4\gamma^4\frac{1}{\rho^2}
\end{aligned} \tag{9.49}$$

using that the radius $\rho = c\beta/\omega$ for motion in a circle. The energy lost in a single revolution is thus $\Delta E = P \times$ the time for a single revolution,

$$\begin{aligned}
\Delta E &= P\frac{2\pi}{\omega} \\
&= P\,2\pi\,\frac{\rho}{c\beta} \\
&= \frac{4\pi}{3}\frac{q^2}{4\pi\epsilon_0}\frac{1}{\rho}\beta^3\gamma^4.
\end{aligned} \tag{9.50}$$

This *synchrotron radiation loss* can be very important for particles in accelerator or storage rings[5]. When ΔE is expressed in GeV and ρ in km, the formula above (with q replaced by $\pm e$—the charge on a proton or an electron) gives

$$\Delta E(\text{GeV}) = 6 \times 10^{-21} \frac{1}{\rho(\text{km})} \gamma^4,$$

since $\beta \approx 1$ for such machines. As we shall see, this is particularly significant for electrons, where the factor γ^4 is $(m_{\text{proton}}/m_{\text{electron}})^4 \approx 11.4 \times 10^{12}$ times larger than for protons with the same energy.

Compare, for example, the synchrotron radiation loss from the protons in the LHC (the Large Hadron Collider at CERN, which stradles the Swiss-French border at Geneva,) with that from the electrons or positrons in the now defunct LEP (Large Electron Positron collider.) The LHC is a collider with crossing points between two beams of protons. These circulate in opposite directions inside adjacent evacuated pipes in a circular tunnel with a 27 km circumference. The LEP occupied the same tunnel, but was used to bring into collision a beam of electrons with a beam of positrons circulating in the opposite direction. So both colliders have the same value for the radius ρ, namely, $27/2\pi$ km. The protons in the LHC each have an energy 7 TeV = 7,000 GeV; dividing by the mass of the proton gives $\gamma \approx 7.4 \times 10^3$, so that for the protons in the LHC we have

$$\Delta E \approx 6 \times 10^{-21} \frac{2\pi}{27} (7.4 \times 10^3)^4 \text{ GeV} = 4.2 \times 10^{-6} \text{ GeV}.$$

For the LHC the protons lose a tiny fraction of their energy on each circuit of the collider. There are of course a very large number of protons (with two beams, each with around 2,800 bunches of protons, each with some 10^{11} protons per bunch)! There results a power loss through synchrotron radiation of around 5 kW = 5 kJ/s. Although 5 kJ is small compared with the stored energy in the beams, it cannot be ignored because it is absorbed by the beam pipes, which have to be maintained at the cryogenic temperature needed for the operation of the superconducting magnets used to hold the beams in their circular orbits. They have to provide magnetic fields of more than 8 T, which would require an unrealistic power consumption with "warm" magnets.

However, for the LEP, where at its peak performance the electrons and positrons (with mass $mc^2 \approx 0.5$ MeV) had energies a little over 100 GeV, giving $\gamma \approx 2 \times 10^5$, we have $\Delta E \approx 2$ GeV. So they lost some 2% of their energy in each circuit through synchrotron radiation, which had to be replaced by the RF power in accelerating modules. On the other hand, because the beams were not as "stiff" as those in the LHC, the magnets needed to provide only 0.135 T and could operate at ambient temperatures, unlike the superconducting magnets of the LHC.

---

[5]For synchrotron light sources, it is not a loss, but their *raison d'être*. The intense synchrotron radiation from giant machines like the Diamond Light Source near Oxford in the UK, or the Advanced Light Source at Berkeley in California, is used in many research applications.

9.6   Radiation from Linear Accelerators

The energy loss through synchrotron radiation militates against the use of circular devices such as the LEP for electrons or positrons at energies above the 100 GeV attained there. So a favored alternative is to use a *linear accelerator*; indeed linear accelerators are used for other purposes, for example as part of the chain of devices used in the initial acceleration of the beam prior to injection into the storage ring of a high-energy accelerator or storage ring. Since the particles undergo acceleration along the linear accelerator, they will emit electromagnetic radiation. This is related to *Bremsstrahlung* ("braking radiation") which is emitted when charged particles are slowed down, for example in collisions—but here their velocity $c\boldsymbol{\beta}$ increases.

Since we are considering what happens in a linear accelerator, the acceleration $\mathbf{a} = c\dot{\boldsymbol{\beta}}$ is parallel to the velocity. Keeping only the dipole term in the expression given in Eq. (9.28) for the power radiated in the direction $\mathbf{n}$, with θ as the angle between $\mathbf{n}$ and the direction of the beam, we find

$$\frac{dP(\theta)}{d\Omega} = \frac{q^2 a^2}{16\pi^2\epsilon_0 c^3} \frac{\sin^2\theta}{(1 - \beta\cos\theta)^5}. \tag{9.51}$$

Integrating over angles gives the total power emitted per particle as

$$P = \frac{q^2 a^2 \gamma^6}{6\pi\epsilon_0 c^3}.$$

In order to assess the significance of this, it is necessary to compare it to the power required to accelerate the particle. To this end, we can rewrite the formula for the power emitted as

$$P = \frac{q^2}{6\pi\epsilon_0 m^2 c^3} \left(\frac{dp}{dt}\right)^2,$$

since the rate of change of the momentum of the particle is

$$\frac{dp}{dt} = mc\gamma^3 a.$$

But this rate of change in momentum is produced by the external force[6] on the particle, which in turn can be expressed using the change in energy of the particle per unit distance along the accelerator,

$$\frac{dp}{dt} = \frac{dE}{dx}.$$

So the interesting quantity is the ratio of the power radiated to the power

---

[6]In fact produced by the RF fields in the klystrons that provide the traveling wave that accelerates the particles.

provided by the external force, namely,

$$
\begin{aligned}
\frac{P}{\frac{dE}{dt}} &= \frac{P}{\frac{dE}{dx}} \frac{1}{c\beta} \\
&= \frac{q^2}{6\pi\epsilon_0 m^2 c^3} \left(\frac{dE}{dx}\right) \frac{1}{c\beta} \\
&\approx \left(\frac{e^2}{6\pi\epsilon_0 m_e c^2}\right) \frac{1}{m_e c^2} \left(\frac{dE}{dx}\right).
\end{aligned}
\tag{9.52}
$$

At the last step we have taken $\beta \approx 1$, and used $q = e$, and $m = m_e$ since we are mainly concerned with electron linear accelerators. In that case, the first term $\left(\frac{e^2}{6\pi\epsilon_0 m_e c^2}\right) = 1.3 \times 10^{-15}$ m. Since even the most ambitious propsals for future high-energy linear accelerators have energy gains around 20–40 MeV/m, the ratio of radiated power to that provided by the external force is utterly negligible.

$$
(1.3 \times 10^{-15} \text{ m}) \times (0.5 \text{ MeV})^{-1} \times 40 \text{ MeV/m} \ll 1!
$$

9.7   Radiation from an Antenna

We will consider as a further example the radiation field from a center-fed linear antenna. This may be idealized as two thin rods each of length $d/2$ placed end to end, say along the z-axis, with a small gap between them. So the rods are at $x = 0, y = 0, \epsilon < |z| < d/2$ with $\epsilon \ll d$. The antenna is "fed" with an alternating current of frequency ω, by a pair of wires, one to each of the rods; these wires lie close to one another so that the net radiation from them is negligible. The current density in the antenna may then be represented by

$$
\mathbf{j}(\mathbf{x}, t) = I \sin\left(\frac{kd}{2} - k|z|\right) \delta(x)\delta(y)\hat{\mathbf{z}}e^{-i\omega t}.
\tag{9.53}
$$

From our previous results, the radiation in the far zone is obtained from

$$
\mathbf{A}(\mathbf{x}) = \frac{\mu_0}{4\pi} \int d^3x' \, \mathbf{j}(\mathbf{x}') \frac{e^{ik|\mathbf{x}-\mathbf{x}'|}}{|\mathbf{x} - \mathbf{x}'|}
$$

with

$$
\frac{e^{ik|\mathbf{x}-\mathbf{x}'|}}{|\mathbf{x} - \mathbf{x}'|} \approx \frac{e^{ikr}}{r} e^{-ik\mathbf{n}\cdot\mathbf{x}'}.
$$

Writing $\mathbf{n} \cdot \mathbf{x}' = z' \cos\theta$ and introducing the expression above for the current density leads to

$$
\begin{aligned}
\mathbf{A} &= \hat{\mathbf{z}}\frac{\mu_0}{4\pi} I \frac{e^{ikr}}{r} \int_{-d/2}^{d/2} dz' \sin\left(\frac{kd}{2} - k|z'|\right) e^{-ikz' \cos\theta} \\
&= \hat{\mathbf{z}}\frac{\mu_0}{4\pi} I \frac{e^{ikr}}{r} \frac{2}{k} \left[\frac{\cos\left(\frac{kd}{2}\cos\theta\right) - \cos\left(\frac{kd}{2}\right)}{\sin^2\theta}\right].
\end{aligned}
\tag{9.54}
$$

Using $\mathbf{B} \approx ik\mathbf{n} \times \mathbf{A} \Rightarrow |\mathbf{B}| = k\sin\theta|\mathbf{A}| = |\mathbf{E}|/c$, we find for the time-averaged power radiated

$$\left\langle \frac{dP}{d\Omega} \right\rangle = \frac{1}{2}\frac{1}{\mu_0 c}|r\mathbf{E}|^2$$
$$= \frac{1}{2}\frac{1}{\mu_0 c}|kcr\sin\theta\mathbf{A}|^2$$
$$= \frac{\mu_0 c}{8\pi^2}I^2\left[\frac{\cos\left(\frac{kd}{2}\cos\theta\right) - \cos\left(\frac{kd}{2}\right)}{\sin\theta}\right]^2. \tag{9.55}$$

If the antenna is *short*, $kd \ll 1$, this reduces to

$$\left\langle \frac{dP}{d\Omega} \right\rangle = \frac{\mu_0 c}{512\pi^2}I^2(kd)^4\sin^2\theta,$$

which has the by now familiar dipole angular distribution and fourth-power dependence on $\omega = ck$. On the other hand, for a *half-wave antenna*, with $kd = \pi$,

$$\left\langle \frac{dP}{d\Omega} \right\rangle = \frac{\mu_0 c}{8\pi^2}I^2\frac{\cos^2\left(\frac{\pi}{2}\cos\theta\right)}{\sin^2\theta},$$

or for a *full-wave antenna*, with $kd = 2\pi$, the result is

$$\left\langle \frac{dP}{d\Omega} \right\rangle = \frac{\mu_0 c}{2\pi^2}I^2\frac{\cos^4\left(\frac{\pi}{2}\cos\theta\right)}{\sin^2\theta}.$$

The angular distributions are thereby changed; more subtle changes in the radiation pattern can be achieved, for example, by *phased arrays* of antennae.

9.8 Exercises for Chapter 9

1 A neutron star rotates with angular rotation frequency ω. It has a magnetic dipole moment of magnitude m, but this is misaligned with the axis of rotation by a constant angle α. Show that it radiates energy at a rate

$$\frac{dE}{dt} = -\frac{\mu_0}{6\pi}\frac{\omega^4}{c^3}m^2\sin^2\alpha.$$

2 Two identical full-wave antennae each of length d are held parallel to one another like the vertical bars in a letter **H**; they are separated from one another by an insulating bar of length $d/2$. The AC feeds to them are identical, except for an adjustable phase difference. Find how the angular distribution of the radiation from this simple array depends on the phase difference.

3 Before the advent of quantum physics, the stability of atoms was puzzling and problematic. Consider a simple model for a hydrogen atom, with an electron (mass $m = 9.1 \times 10^{-31}$ kg, charge $-e = -1.6 \times 10^{-19}$ C), orbiting a proton with charge $+e$ that may be regarded as stationary. Suppose it moves in a circular orbit af radius 5.3×10^{-11} m. Because of its circular motion, the electron will radiate. Assuming that the angular momentum of the electron is conserved, show that the decrease in the energy of the atom implies an increase in the angular velocity of the electron and a decrease in the radius of its orbit. Estimate how long it would take for the radius of the orbit to be reduced to one-half of its original value.

Chapter 10

Media

In most of the foregoing chapters we have considered fields *in vacuo*, where electromagnetic radiation propagates without dispersion—that is to say that the speed of light is the same for all frequencies. But in a medium this is no longer the case.

So we will start again from Maxwell's equations, but now in their form appropriate to fields in a medium. The *homogeneous* equations

$$\nabla \cdot \mathbf{B} = 0, \qquad \nabla \times \mathbf{E} = -\dot{\mathbf{B}} \tag{10.1}$$

remain unchanged, and allow us to introduce the potentials $\mathbf{A}$ and Φ as before, in terms of which we still have

$$\mathbf{B} = \nabla \times \mathbf{A}, \qquad \mathbf{E} = -\dot{\mathbf{A}} - \nabla\Phi. \tag{10.2}$$

The *inhomogeneous* equations

$$\nabla \cdot \mathbf{D} = \rho, \qquad \nabla \times \mathbf{H} = \mathbf{j} + \dot{\mathbf{D}} \tag{10.3}$$

introduce the fields $\mathbf{D}$ and $\mathbf{H}$, which are essentially phenomenological, macroscopic fields. They are related to the more fundamental fields $\mathbf{E}$ and $\mathbf{B}$ by the *constitutive relations*:

$$\mathbf{D} = \epsilon\mathbf{E}, \qquad \mathbf{B} = \mu\mathbf{H}. \tag{10.4}$$

(In a conductor we also have $\mathbf{j} = \sigma\mathbf{E}$.) Although we have written the permittivity ϵ, the permeability μ, and the conductivity σ as scalar quantities, they are more generally tensors in an anisotropic medium. We will for the present suppose them to be constants. As before, we introduce the polarization $\mathbf{P}$ and the magnetization $\mathbf{M}$ by

$$\mathbf{D} - \epsilon_0\mathbf{E} = \mathbf{P}, \qquad \mathbf{B} = \mu_0(\mathbf{H} + \mathbf{M}). \tag{10.5}$$

The relative permittivity ϵ_r and the relative permeability μ_r are defined[1] using the vacuum constants ϵ_0, μ_0 by

$$\epsilon = \epsilon_0\epsilon_r, \qquad \mu = \mu_0\mu_r. \tag{10.6}$$

---

[1] Evidently as dimensionless quantities!

Direct substitution allows one to conclude that

$$\left(\epsilon\mu\frac{\partial^2}{\partial t^2} - \nabla^2\right)\Phi = \frac{\rho}{\epsilon}, \qquad \left(\epsilon\mu\frac{\partial^2}{\partial t^2} - \nabla^2\right)\mathbf{A} = \mu\mathbf{j}, \qquad (10.7)$$

where we have used the freedom to choose a gauge in which

$$\nabla \cdot \mathbf{A} + \epsilon\mu\dot{\Phi} = 0.$$

In the absence of sources, $\mathbf{j} = 0$, $\rho = 0$, these have solutions of the form $\exp[-i(\omega t - \mathbf{k} \cdot \mathbf{x})]$ of propagating waves, with *phase velocity* $v = \omega/k$ given by

$$v^2 = \frac{1}{\epsilon\mu}. \qquad (10.8)$$

Of course, in vacuum we have $c^2 = \frac{1}{\epsilon_0\mu_0}$. Note that when there are no free charges $(\rho = 0)$, we may take $\Phi = 0$, so that we are in radiation gauge, when also $\nabla \cdot \mathbf{A} = 0$. Then we have

$$\mathbf{E} = -\dot{\mathbf{A}}, \qquad \mathbf{B} = \nabla \times \mathbf{A}. \qquad (10.9)$$

For plane waves with frequency ω, it follows that

$$\mathbf{E} = i\omega\mathbf{A}, \qquad \mathbf{B} = i\mathbf{k} \times \mathbf{A} = \frac{\mathbf{k} \times \mathbf{E}}{\omega}, \qquad (10.10)$$

and using the gauge condition $\nabla \cdot \mathbf{A} = 0 \Rightarrow \mathbf{k} \cdot \mathbf{E} = 0$, we have illustrated the fact that $\mathbf{k}, \mathbf{E}, \mathbf{B}$ are mutually orthogonal vectors: electromagnetic waves are *transverse*. In any case, the $\mathbf{E}$ and $\mathbf{B}$ fields each satisfy the same wave equation as $\mathbf{A}$, namely,[2]

$$\nabla^2\mathbf{B} = \frac{1}{v^2}\frac{\partial^2\mathbf{B}}{\partial t^2}, \qquad \nabla^2\mathbf{E} = \frac{1}{v^2}\frac{\partial^2\mathbf{E}}{\partial t^2}. \qquad (10.12)$$

10.1 Dispersion

The relative permittivity ϵ_r and permeability μ_r may (and usually do) depend on the frequency ω, which means that different frequencies propagate at different speeds. This leads to *dispersion*, since the ratio c/v is what in optics is called the *refractive index* of the medium. At optical frequencies the relative permeability is to good approximation equal to 1, so that the refractive index n is given by $n \approx \sqrt{\epsilon_r}$.

---

[2]In conductors, the inclusion of $\mathbf{j} = \sigma\mathbf{E}$ modifies these equations to give

$$\nabla^2\mathbf{B} = \frac{1}{v^2}\frac{\partial^2\mathbf{B}}{\partial t^2} + \sigma\mu\frac{\partial\mathbf{B}}{\partial t}, \qquad \nabla^2\mathbf{E} = \frac{1}{v^2}\frac{\partial^2\mathbf{E}}{\partial t^2} + \sigma\mu\frac{\partial\mathbf{E}}{\partial t}. \qquad (10.11)$$

For waves of frequency ω, the result of this is to replace $v^{-2} = \epsilon\mu$ by $\epsilon\mu + i\sigma\mu/\omega$, so in effect to add an imaginary part σ/ω to the permittivity.

10.1.1   Newton on the "Phænomena of Colours"

In the beginning of the year 1666 (at which time I applyed my self to the grinding of Optick glasses of other figures than Spherical,) I procured me a Triangular glasse Prisme to try therewith the celebrated *Phænomena of Colours.* And in order thereto having darkened my chamber, and made a small hole in my window-shuts, to let in a convenient quantity of the Suns light, I placed my Prisme at his entrance, that it might be thereby refracted to the opposite wall. It was at first a very pleasing divertisement to view the vivid & intense colours produced thereby; but after a while applying my self to consider them more circumspectly, I became surprized to see them in an *oblong* form, which according to the received laws of Refraction, I expected should have been *circular.*[3]

This vivid demonstration that the deviation of light by refraction on passing through the prism depends on the color of the light was followed with meticulous care by Newton in a series of other experiments. The white light from the sun is a mixture of different colors—all the colors of the rainbow, as Newton recorded: red, orange, yellow, green, blue, indigo, violet. Although he was himself an advocate of a corpuscular theory of light, he described phenomena that are most readily interpreted in the wave theory, notably what are still called "Newton's rings" and other manifestations of interference. He even measured what would now be recognized as the wavelength of light and found it to depend on the colour.

The "received laws of Refraction" include Snell's law.[4] They may be derived from the equations of electromagnetism, as we will now show.

10.2   Refraction

When a ray of light passes from one transparent medium to another, its path is deviated. If we recognise that the "rays" of geometrical optics are orthogonal to the wave-fronts of physical optics, this means that a plane wave with wave-vector $\mathbf{k}$ incident on a planar interface between two media will lead to a transmitted plane wave on the opposite side of the interface with wave-vector $\mathbf{k}''$. There will also be a reflected wave on the same side as the incident wave with wave-vector $\mathbf{k}'$. What we have to do is to find how these three wave-vectors are related to one another. And to do this, we will need to determine the consequences of Maxwell's laws at the interface; these are the *boundary conditions* that apply at the interfacial boundary between the two media.

Let us suppose that the interface is the y-z plane, and the incident wave has $k_z = 0$ (so its wave fronts are perpendicular to the x-y plane, or otherwise stated, the incident rays are in the x-y plane). We shall also suppose that the incident wave is on the side $x < 0$ of the interface, as is the reflected wave,

---

[3]Isaac Newton, *Phil. Trans. R. Soc.* (1671).

[4]Stated by Snell in 1621, published by Descartes in his *Discourse on Method,* but already given by Ibn Sahl before 1000.

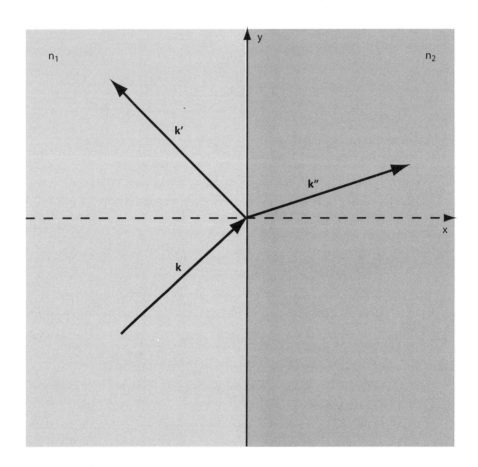

Figure 10.1. The interface between two media.

and that for $x < 0$ the refractive index is n_1. On the other side, in which the transmitted wave propagates, the refractive index is n_2. (Fig. 10.1) The reflected and refracted wave will have the same time dependence, $e^{-i\omega t}$, as the incident wave, so that for the three waves, incident, reflected, and transmitted, we have, respectively:

$$\mathbf{E} = \mathbf{E}_0 \exp\left(i\mathbf{k} \cdot \mathbf{x} - i\omega t\right), \qquad \mathbf{B} = \frac{\mathbf{k} \times \mathbf{E}}{\omega}, \qquad (10.13)$$

$$\mathbf{E}' = \mathbf{E}_0' \exp\left(i\mathbf{k}' \cdot \mathbf{x} - i\omega t\right), \qquad \mathbf{B}' = \frac{\mathbf{k}' \times \mathbf{E}'}{\omega}, \qquad (10.14)$$

$$\mathbf{E}'' = \mathbf{E}_0'' \exp\left(i\mathbf{k}'' \cdot \mathbf{x} - i\omega t\right), \qquad \mathbf{B}'' = \frac{\mathbf{k}'' \times \mathbf{E}''}{\omega}. \qquad (10.15)$$

10.2.1   The Boundary Conditions at the Interface

We need to go beyond the results of *geometrical*, or *ray optics*, which are in effect those of the kinematics of refraction and reflection. To take account also of the wave nature of electromagnetic radiation, and indeed of light, we need to consider the conditions on the fields that hold at the interface between the media, namely, the plane $x = 0$.

The differential form of Maxwell's equations that we have used so far can be integrated to give

$$\nabla \cdot \mathbf{B} = 0 \Longrightarrow \iint_{\partial V} \mathbf{B} \cdot d\mathbf{S} = 0,$$

$$\nabla \times \mathbf{E} = -\dot{\mathbf{B}} \Longrightarrow \oint_{\partial S} \mathbf{E} \cdot d\mathbf{l} = -\frac{\partial}{\partial t} \iint_S \mathbf{B} \cdot d\mathbf{S},$$

$$\nabla \cdot \mathbf{D} = \rho \Longrightarrow \iint_{\partial V} \mathbf{D} \cdot d\mathbf{S} = \iiint_V \rho \, dV,$$

$$\nabla \times \mathbf{H} = \mathbf{j} + \dot{\mathbf{D}} \Longrightarrow \oint_{\partial S} \mathbf{H} \cdot d\mathbf{l} = \iint_S \mathbf{j} \cdot d\mathbf{S} + \frac{\partial}{\partial t} \iint_S \mathbf{D} \cdot d\mathbf{S}. \qquad (10.16)$$

In these equations, ∂V is a closed surface surrounding a volume V and ∂S is a closed contour that is the boundary of a surface S; $d\mathbf{S} = \mathbf{n}dS$ is a surface element with normal $\mathbf{n}$ and $d\mathbf{l} = \mathbf{t}dl$ is a line element with $\mathbf{t}$ the unit vector in the direction of the tangent to the contour.

By suitable choice of the volume V (cylindrical, with end surfaces on opposite sides of the boundary) and of the contour ∂S (a narrow rectangle with long edges on opposite sides of the boundary) one can derive continuity conditions that hold at the boundary $x = 0$:

   D_x and B_x, the normal components of $\mathbf{D}$ and $\mathbf{B}$ are continuous
   across the inteface $x = 0$; and likewise E_y, E_z, H_y, H_z, the tangential
   components of $\mathbf{E}$ and $\mathbf{H}$ are continuous.

On the interface $x = 0$ the boundary conditions have to hold at all times, which means that the phase factors of the three waves (Eqs (10.13), (10.14), and (10.15)) must be equal to one another, so that we have

$$(\mathbf{k} \cdot \mathbf{x})_{x=0} = (\mathbf{k}' \cdot \mathbf{x})_{x=0} = (\mathbf{k}'' \cdot \mathbf{x})_{x=0}. \qquad (10.17)$$

This implies that the three wave-vectors $\mathbf{k}, \mathbf{k}', \mathbf{k}''$ are coplanar, which is one of the "received laws of Refraction." Furthermore, with θ_i as the angle between the incident direction $\mathbf{k}$ and the normal to the interface, $(\mathbf{k} \cdot \mathbf{x})_{x=0} = k \sin \theta_i$. Likewise with θ_r the angle between the reflected rays and the normal, and θ_t that for the transmitted rays, we have $(\mathbf{k}' \cdot \mathbf{x})_{x=0} = k' \sin \theta_r = k \sin \theta_r$ and $(\mathbf{k}'' \cdot \mathbf{x})_{x=0} = k'' \sin \theta_t = \frac{n_2}{n_1} k \sin \theta_t$. (We have used $|\mathbf{k}| = |\mathbf{k}_r| = k$, since both he incident and reflected waves are in the same region, $x < 0$, while the transmitted wave, lying in the region $x > 0$, has wave-number that differs by the factor $\frac{n_2}{n_1}$.) So we recover Snell's law for refraction,

$$n_1 \sin \theta_i = n_2 \sin \theta_t, \tag{10.18}$$

and also the law of reflection,

$$\theta_r = \theta_i. \tag{10.19}$$

Note that if $n_1 > n_2$, and $\sin \theta_i > n_2/n_i$, it is impossible to satisfy the condition imposed by Snell's law. There is no transmitted wave, and the incident wave is totally reflected—total internal reflection.[5]

To derive further consequences of the boundary conditions, it is simplest to consider them separately for the two independent cases; one with the polarization of the incident radiation (specified by the direction of the electric field vector!) perpendicular to the plane of incidence, i.e. the plane defined by $\mathbf{k}$ and $\mathbf{n}$, the normal to the interface; the other with the polarization of the incident radiation parallel to the plane of incidence. The general case can then be treated by a suitable superposition of these two cases.

10.2.1.1 Polarization Perpendicular to Plane of Incidence

The consequences of the boundary conditions are that the tangential components of $\mathbf{E} + \mathbf{E}'' - \mathbf{E}'$ should vanish, and likewise the tangential components of $\frac{1}{\mu_1}(\mathbf{k} \times \mathbf{E} + \mathbf{k}'' \times \mathbf{E}'') - \frac{1}{\mu_2}\mathbf{k}' \times \mathbf{E}'$ also should vanish. These in turn lead to

$$E_0 + E_0'' = E_0',$$

$$\sqrt{\frac{\epsilon_1}{\mu_1}}(E_0 - E_0'') \cos \theta_i = \sqrt{\frac{\epsilon_2}{\mu_2}} E_0' \cos \theta_t. \tag{10.20}$$

At optical frequencies it is usually an excellent approximation to set $\mu_1 = \mu_2$ ($= \mu_0$), and thus $\sqrt{\frac{\epsilon_1}{\mu_1}} = n_1$ and $\sqrt{\frac{\epsilon_2}{\mu_2}} = n_2$. For simplicity, we will make this replacement, and then obtain

$$\frac{E_0'}{E_0} = \frac{2n_1 \cos \theta_i}{n_1 \cos \theta_i + n_2 \cos \theta_t},$$

$$\frac{E_0''}{E_0} = \frac{n_1 \cos \theta_i - n_2 \cos \theta_t}{n_1 \cos \theta_i + n_2 \cos \theta_t}. \tag{10.21}$$

---

[5] A more careful analysis of what happens at the interface in this situation confirms that there is no progressing wave on the less optically dense side of the inteface, but there is an *evanescent wave*, which instead of having an oscillating dependence on the distance from the inteface falls away exponentially.

For plane waves in a medium, the Poynting vector is given by

$$\mathbf{S} = \frac{1}{\mu}\mathbf{E} \times \mathbf{B} = \frac{1}{\mu\omega}\frac{n}{ck}E^2\mathbf{k},$$

with $k = |\mathbf{k}|$. So the rate at which energy in the incident beam strikes an area A of the interface is

$$I = \frac{An_1\cos\theta_i}{c\mu\omega}E_0^2,$$

while the rates at which energy leaves that area in the transmitted and reflected beams are, respectively,

$$I' = \frac{An_2\cos\theta_t}{c\mu\omega}E_0'^2,$$

$$I'' = \frac{An_1\cos\theta_i}{c\mu\omega}E_0''^2.$$

The *transmission coefficient* is defined as the ratio $T = I'/I$ and likewise the *reflection coefficient* is $R = I''/I$. So we have in this case of polarization perpendicular to the plane of incidence (indicated by the suffix $\perp$ on the coefficients)

$$T_\perp = \frac{4n_1n_2\cos\theta_i\cos\theta_t}{(n_1\cos\theta_i + n_2\cos\theta_t)^2}, \tag{10.22}$$

$$R_\perp = \frac{(n_1\cos\theta_i - n_2\cos\theta_t)^2}{(n_1\cos\theta_i + n_2\cos\theta_t)^2}. \tag{10.23}$$

It is gratifying to find that $T_\perp + R_\perp = 1$ as expected.

Note that both the transmitted and reflected radiation is polarized, still perpendicular to the plane of incidence.

10.2.1.2 Polarization Parallel to Plane of Incidence

We now find that the boundary conditions require

$$(E_0 - E_0'')\cos\theta_i = E_0'\cos\theta_t,$$

$$\sqrt{\frac{\epsilon_1}{\mu_1}}(E_0 - E_0'') = \sqrt{\frac{\epsilon_2}{\mu_2}}E_0'. \tag{10.24}$$

It follows that the coefficients $T_\parallel$ and $R_\parallel$ with this incident polarization are

$$T_\parallel = \frac{4n_1n_2\cos\theta_i\cos\theta_t}{(n_1\cos\theta_t + n_2\cos\theta_i)^2}, \tag{10.25}$$

$$R_\parallel = \frac{(n_1\cos\theta_t - n_2\cos\theta_i)^2}{(n_1\cos\theta_t + n_2\cos\theta_i)^2}. \tag{10.26}$$

Once again, the transmitted and reflected radiation is polarized, but this time parallel to the plane of incidence. And once again the coefficients add to unity, $T_\parallel + R_\parallel = 1$.

10.2.1.3 Brewster Angle

An unpolarized incident wave is an incoherent mixture of the two polarizations just considered. But it is apparent from the results obtained that the resultant reflected and transmitted radiation is partially polarized. Indeed there is an angle of incidence for light polarized parallel to the incident plane, the *Brewster angle*, at which the reflected wave vanishes. This means that at that angle of incidence, the reflected wave for any incident polarization is fully polarized perpendicular to the incident plane.[6] The Brewster angle θ_B is the angle of incidence θ_i for which $\frac{\mu_1}{\mu_2} n_2 \cos\theta_i - n_1 \cos\theta_t = 0$ is satisfied, or, setting $\mu_1 = \mu_2$ and using also Snell's law,

$$\tan\theta_B = \frac{n_2}{n_1}.$$

The partial polarization at other incident angles, especially those close to the Brewster angle, means that a polarizing filter that transmits only with polarization in the vertical direction will cut down the glare from reflections at horizontal surfaces, such as the sea, an effect used by photographers. It is also exploited in polarizing sunglasses for the same reason, since most of the dazzle in sunny situations is from reflections at horizontal surfaces.

10.3   Čerenkov Radiation

Because the speed of light in a medium with refractive index n is less than the speed of light *in vacuo*, it is possible for a charged particle to move through such a medium at a speed greater than the speed of light in the medium. It then emits *Čerenkov radiation*,[7] which has an analogy with the shock waves (for example the sonic boom) produced by an object moving faster than the speed of sound. We will give a simplified explanation of the theory, following that of Frank and Tamm.[8]

We assume that the particle has charge q and moves with constant speed $v = c\beta$ along the z-axis. So we ignore multiple scattering, ionization, and so on, and also ignore dispersion, and take $\mu_r = 1$. Radiation reaction and absorption of the radiation are also neglected. The electromagnetic field produced by the particle may then be regarded as the result of spherical waves of retarded potential, continuously emitted by the particle, and propagating with speed c/n. These will be in phase only along the direction making an angle θ with the z-axis, (Fig 10.2) where

$$\cos\theta = \frac{1}{\beta n}. \tag{10.27}$$

Note that this condition requires $\beta n > 1$. In any other direction, successive waves will interfere, and there is no resultant radiation. But if $\beta n > 1$ there will be radiation in the direction θ. This is Čerenkov radiation.

---

[6]Polarization by reflection had been observed in 1808 by Etienne-Louis Malus, but the complete polarization at a critical angle was first described in 1815 by Sir David Brewster (1781-1868).

[7]Named for Pavel Alekseyevich Čerenkov (1904–1990), who discovered it in 1934.

[8]Ilya Mikhailovich Frank (1908–1990) and Igor Yevgenyevich Tamm (1895–1971).

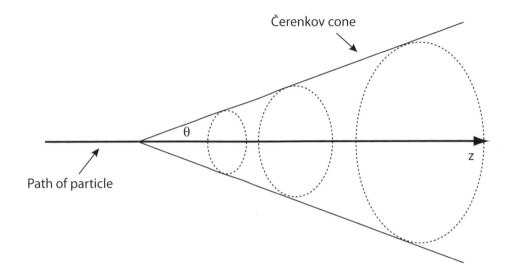

Figure 10.2. The Čerenkov cone.

We now give the outline of a more detailed derivation, starting as ever from Maxwell's equations. From the definition of the electric polarization

$$\mathbf{P} = \mathbf{D} - \epsilon_0 \mathbf{E} = (\epsilon - \epsilon_0)\mathbf{E}$$

and making the usual Fourier expansion to go from time to frequency, we obtain

$$\mathbf{P}(\omega) = (n^2 - 1)\mathbf{E}(\omega),$$

where n is the refractive index at frequency ω. The corresponding frequency component of the vector potential $\mathbf{A}$ will satisfy

$$\nabla^2 \mathbf{A}(\omega) + \frac{\omega^2 n^2}{c^2} \mathbf{A}(\omega) = -\mu_0 \mathbf{j}(\omega).$$

The current density is that due to the particle, so we have

$$j_x = j_y = 0, \quad j_z = qv\delta(x)\delta(y)\delta(z - vt). \tag{10.28}$$

The Fourier transform from time t to frequency ω gives

$$\mathbf{j}(\omega) = q\exp(-i\omega z/v)\delta(x)\delta(y).$$

Because the problem has cylindrical symmetry, it is appropriate to rewrite this using cylindrical coordinates ρ, ϕ, z, with the result

$$j_z(\omega) = \frac{q}{2\pi\rho} \exp(-i\omega z/v)\delta(\rho). \tag{10.29}$$

The vector potential, now in cylindrical polar coordinates, satisfies

$$A_\rho = A_\phi = 0, \quad A_z(\omega) = u(\rho)\exp(-i\omega z/v), \tag{10.30}$$

where the function $u(\rho)$ satisfies

$$\frac{\partial^2 u}{\partial \rho^2} + \frac{1}{\rho}\frac{\partial u}{\partial \rho} + s^2 u = -\frac{2q}{c\rho}\delta(\rho). \tag{10.31}$$

The parameter s^2 is defined by

$$s^2 = \frac{\omega^2}{v^2}(\beta^2 n^2 - 1),$$

which is positive when $\beta n > 1$. Everywhere except at $\rho = 0$, the equation for u is in fact a Bessel equation. To obtain the correct singularity so as to match the behavior as $\rho \to 0$ requires

$$\lim_{\rho \to 0} \rho \frac{\partial u}{\partial \rho} = -2qc,$$

which together with the requirement that u leads only to outgoing waves identifies the solution to be

$$u = -\frac{i\pi q}{c} H_0^{(2)}(s\rho),$$

where H_0^2 is a Hankel function. The explicit form of this function is for our purpose less interesting than its asymptotic behavior at large ρ. This follows from

$$H_0^{(2)}(w) \sim \sqrt{2/(\pi w)}\exp(-i(w - \pi/4));$$

so introducing the $\exp(i\omega t)$ factor into $A_z(\omega)$ we obtain

$$A_z(\omega)\exp(i\omega t) \sim -\frac{q}{c\sqrt{2\pi s\rho}}\exp\left[i\omega\left(t - \frac{z}{v}\right) - i\left(s\rho - \frac{3\pi}{4}\right)\right]$$

$$= -\frac{q}{c\sqrt{2\pi s\rho}}\exp\left[i\omega\left(t - \frac{z\cos\theta + \rho\sin\theta}{c/n}\right) + \frac{3}{4}i\pi\right]. \quad (10.32)$$

This can be seen to be a wave propagating to infinity along the direction θ. Further calculation shows that the spectrum of the radiation is continuous as a function of frequency, and that the relative intensity is essentially proportional to the frequency, which means that the visible light emitted is dominated by the higher-frequency range and appears as a brilliant blue. Čerenkov radiation is responsible for the blue glow seen around fuel rods in a water-moderated nuclear reactor.

Čerenkov detectors are used in high-energy particle physics experiments. They respond to the radiation and, because of the dependence on the mass of the threshold energy for Čerenkov radiation, they can, at a given energy, discriminate between lighter particles that radiate and more massive particles (which move more slowly) that do not. Furthermore, the opening angle of the cone irradiated can be used directly to determine the speed of the particle.[9]

10.4   Exercises for Chapter 10

1   A glass prism (refractive index $n = 1.5$) having triangular cross-section with edges 3, 4, and 5 cm wide is held with its 3-cm-wide face horizontal, parallel to a plane glass sheet, but separated from it by 0.01 mm (Fig. 10.3). A beam of red light (wave length 650 nm) from a laser is directed at normal incidence through the 5-cm-wide face so as to strike the horizontal face at an angle θ, with $\sin\theta = 0.8$. Show that the light suffers total internal reflection. However, a much fainter beam is seen to enter the glass sheet, despite its being separated from the prism. Explain this phenomenon. Estimate by how much the intensity of the beam is diminished in crossing the gap between the prism and the glass sheet.

2   The lenses in optical devices such as cameras are often *bloomed*, that is, coated with a thin layer of a transparent material chosen so as to minimize reflections from the surfaces of the lens. Explain how this is effective, and why, although reflections are not entirely eliminated, the reflected light appears to be purple. How is the thickness and refractive index of the coating chosen?

---

[9]Super-Kamiokande is a 50,000 tonne water Čerenkov detector at the Kamioka Observatory in Japan. It is used to explore the physics of neutrinos through their interactions, which can produce electrons or muons. These generate flashes of Čerenkov radiation that are detected with 11,000 photomultiplier tubes.

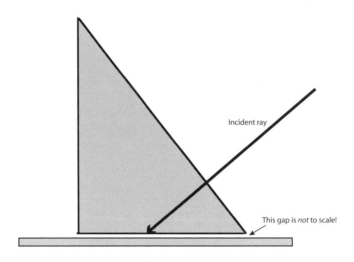

Figure 10.3. A triangular glass prism supported above a glass plate.

3 A spark discharge is able to produce a pulse of white light of duration 50 ns. When this pulse is passed through a block of glass that is several meters long, the length of the pulse (that is, the distance between its leading and trailing edges, as well as its duration) is observed to have increased. Explain this phenomenon and relate it to the spectrum of colors observed when a beam of white light passes through a prism.

Chapter 11

Scattering

We turn now to the consideration of the problem of scattering. In a typical scattering problem, the scatterer will be localized in some region of space. At some early time we are given some (free) fields A_{in}^{μ} which propagate forward in time, and then interact with the scatterer, *inducing* in it an electric polarization **P** and magnetization **M** with time dependence driven by that of the incident fields. These now act as source terms for the potentials.[1] We want to find how an incident plane wave is scattered, and in particular the angular dependence of the intensity at large distances from the scattering region.

So we have

$$\Box A_{in}^{\mu} = 0,$$

since the incident fields are free, and

$$\Box A^{\mu} = \mu_0 j^{\mu},$$

in which the fields A^{μ} are a superposition

$$A^{\mu} = A_{in}^{\mu} + A_{sc}^{\mu}$$

of the incident fields and the scattered fields. The incident fields are in effect the complementary function for the inhomogeneous equation that is the field equation (i.e., a solution to the corresponding homogeneous equation), while the scattered fields are a particular integral, which physical considerations dictate should be the one with only outgoing waves, and so are determined by the retarded potentials

$$A_{sc}^{\mu} = \frac{\mu_0}{4\pi} \int d^3x' \, j^{\mu}(\mathbf{x}', t - r/c)\frac{1}{r},$$

giving

$$A^{\mu} = A_{in}^{\mu} + \frac{\mu_0}{4\pi} \int d^3x' \, j^{\mu}(\mathbf{x}', t - r/c)\frac{1}{r}. \tag{11.1}$$

---

[1] A static electric polarization or megnetization, as in ferroelectric media or permanent magnets, will make no contribution to the scattering.

Let us specialize to the case when the *incident* field is a plane wave, with angular frequency ω; the scattered wave will then have the same frequency (at least for the simple cases we shall consider), and is generated by the charge and current distributions induced in the scatterer by the incident radiation—which will oscillate with the same frequency ω.

11.1   Scattering from a Small Scatterer

If the scatterer is small, we may use a multipole expansion, and then in the far field, keeping only the dipole contributions,

$$\mathbf{E}_{\text{sc}} = \frac{k^2}{4\pi\epsilon_0}\frac{e^{ikr}}{r}[(\mathbf{n} \times \mathbf{p}) \times \mathbf{n} - \mathbf{n} \times \mathbf{m}/c], \tag{11.2}$$

$$\mathbf{B}_{\text{sc}} = \frac{1}{c}\mathbf{n} \times \mathbf{E}_{\text{sc}}, \tag{11.3}$$

where $\mathbf{p}$ and $\mathbf{m}$ are the induced electric and magnetic dipole moments of the scatterer.

If the incident plane wave is in the direction of the unit vector $\mathbf{n}_0$, we may write

$$\mathbf{E}_{\text{in}} = \mathbf{E}_0 e^{i\mathbf{k}_0 \cdot \mathbf{x}}, \tag{11.4}$$

$$\mathbf{B}_{\text{in}} = \frac{1}{c}\mathbf{n}_0 \times \mathbf{E}_{\text{in}}. \tag{11.5}$$

The wave vector is $\mathbf{k}_0 = k\mathbf{n}_0$, and the time dependence is in the amplitude $\mathbf{E}_0$, which also specifies the polarization of the incident radiation. The incident *flux*, the power per unit area normal to the incident wave, is, after taking the time average,

$$\langle|\mathbf{S}_{\text{in}}|\rangle = \frac{1}{2\mu_0 c}|\mathbf{E}_{\text{in}}|^2 = \frac{1}{2\mu_0 c}E_0^2. \tag{11.6}$$

Similarly, the flux of energy in the scattered wave is

$$\langle|\mathbf{S}_{\text{sc}}|\rangle = \frac{1}{2\mu_0 c}|\mathbf{E}_{\text{sc}}|^2. \tag{11.7}$$

This flux is radially outward, that is, along the direction $\mathbf{n}$, and falls off as $\frac{1}{r^2}$. The polarization of the scattered radiation is determined by the direction of $\mathbf{E}_{\text{sc}}$.

The intensity of scattered radiation that might be measured by a detector placed at distance r and in the direction $\mathbf{n}$ from the scatterer will depend on the solid angle it subtends at the scatterer. It is characterized by the *differential scattering cross-section*, defined as the power radiated in a direction $\mathbf{n}$ per unit of incident flux (with direction $\mathbf{n}_0$).[2] We can also take account of the polarization

---

[2]Integration of the differential cross-section with respect to $d\Omega$ gives the total cross-section, the rate at which energy is removed by scattering from an incident wave normalized to unit flux.

state of the incident and scattered waves by introducing unit vectors (which may be complex if, for example, the polarization is circular) $\boldsymbol{\epsilon}_0$ and $\boldsymbol{\epsilon}$, respectively. Then

$$
\begin{aligned}
\frac{d\sigma}{d\Omega}(\mathbf{n}, \boldsymbol{\epsilon}; \mathbf{n}_0, \boldsymbol{\epsilon}_0) &= \frac{r^2 |\boldsymbol{\epsilon}^* \cdot \mathbf{E}_{\mathrm{sc}}|^2}{|\boldsymbol{\epsilon}_0^* \cdot \mathbf{E}_{\mathrm{in}}|^2} \\
&= \left(\frac{k^2}{4\pi\epsilon_0}\right)^2 \frac{1}{E_0^2} |\boldsymbol{\epsilon}^* \cdot [(\mathbf{n} \times \mathbf{p}) \times \mathbf{n} - \mathbf{n} \times \mathbf{m}/c]|^2 \\
&= \left(\frac{k^2}{4\pi\epsilon_0}\right)^2 \frac{1}{E_0^2} |\boldsymbol{\epsilon}^* \cdot \mathbf{p} - \boldsymbol{\epsilon}^* \cdot (\mathbf{n} \times \mathbf{m})/c|^2 .
\end{aligned} \tag{11.8}
$$

(We have used $\mathbf{n} \cdot \boldsymbol{\epsilon} = 0$.) Note that unless the dipole moments vanish the radiation scattered has a fourth-power frequency dependence, as contained in the factor k^4. This is *Rayleigh's law*.

11.2 Many Scatterers

In the previous section, we supposed that the scatterer was at the origin. Had it been at the point $\mathbf{x}_j$, the incident wave would have been modified by an additional factor $\exp[ik\mathbf{n}_0 \cdot \mathbf{x}_j]$ in $\mathbf{E}_0$, and the phase at the (distant) point at which the scattered wave is observed would likewise be modified by a phase $\exp[-ik\mathbf{n} \cdot \mathbf{x}_j]$. The resultant phase change overall between the scattered and the incident waves, $\exp[i\mathbf{q} \cdot \mathbf{x}_j]$, where $\mathbf{q} = k(\mathbf{n}_0 - \mathbf{n})$, makes no difference to the resultant differential cross-section.

However, if there are a number of scattering centers, located at different points, the phase differences between them have an important consequence, because of interference effects. For a collection of dipole scatterers, we will now find

$$
\frac{d\sigma}{d\Omega} = \left(\frac{k^2}{4\pi\epsilon_0}\right)^2 \frac{1}{E_0^2} \left| \sum_j \boldsymbol{\epsilon}^* \cdot \left[\mathbf{p}_j - \frac{1}{c}(\mathbf{n} \times \mathbf{m}_j)\right] e^{i\mathbf{q}\cdot\mathbf{x}_j} \right|^2 , \tag{11.9}
$$

which depends sensitively on the positions $\mathbf{x}_j$ of the scatterers. This is the basis, for example, of the efficacy of X-ray scattering to determine crystal structure. If all the scatterers are identical, the effect is to multiply the cross-section for a single scatterer by

$$
\mathcal{F}(\mathbf{q}) = \left| \sum_j e^{i\mathbf{q}\cdot\mathbf{x}_j} \right|^2 = \sum_{ij} e^{i\mathbf{q}\cdot(\mathbf{x}_i - \mathbf{x}_j)}, \tag{11.10}
$$

the *structure factor*. In the forward direction where $\mathbf{q} = 0$, we have $\mathcal{F} = N^2$ (where N is the number of scatterers). When the distribution of scatterers is random the phases of the off-diagonal terms in this sum cancel except very close to the forward direction, and the result is that, for this case of *incoherent* scattering, $\mathcal{F}(\mathbf{q}) = N$ so long as $|\mathbf{q}| \gg a^{-1}$, a being a typical distance between the scatterers.

On the other hand, if the scatterers are arranged in a regular array, as for example in a crystal, the coherence of the contributions to the scattering leads to a characteristic pattern. Thus, for a simple cubic structure, with lattice spacing a, with N_1, N_2, N_3 lattice sites along the three axes of the crystal, and with $\mathbf{q} = (q_1, q_2, q_3)$ when referred to these axes, one finds[3]

$$\mathcal{F}(\mathbf{q}) = \left[\frac{\sin^2\left(\frac{N_1 q_1 a}{2}\right)}{\sin^2\left(\frac{q_1 a}{2}\right)}\right] \cdot \left[\frac{\sin^2\left(\frac{N_2 q_2 a}{2}\right)}{\sin^2\left(\frac{q_2 a}{2}\right)}\right] \cdot \left[\frac{\sin^2\left(\frac{N_3 q_3 a}{2}\right)}{\sin^2\left(\frac{q_3 a}{2}\right)}\right]. \tag{11.11}$$

Each factor peaks sharply around $qa = 0, 2\pi, 4\pi, \ldots$, that is, whenever the Bragg condition is satisfied, and then, just as in the forward direction, $\mathcal{F} = N^2$. The number of peaks is limited by the maximum value that qa can attain, $qa \le 2ka$, so that at long wavelengths only the forward peak occurs. And this has a width determined by $q_i \le 2\pi/N_i a$, corresponding to scattering angles less than or of order λ/L, where L is the linear size of the crystal.

11.3 Scattering from the Sky

The light scattered from a cloudless sky appears to be blue.[4] The explanation of this may be given at various levels of sophistication. It is already apparent from the Rayleigh formula that the k^4 dependence of dipole scattering will lead to a dominance of the shorter wavelengths in the scattered radiation from an originally white source. We are now ready for a more detailed description of the origin of this dipole scattering.

We may suppose that there are no free charges or currents in the sky.[5] This means $\rho = \mathbf{j} = 0$, and our previous approach fails to give any scattering whatever! The reason is that we have been considering sources *in vacuo*, but air is not a vacuum, so that $\epsilon \equiv \epsilon_0 \epsilon_r \ne \epsilon_0$ and $\mu \equiv \mu_0 \mu_r \ne \mu_0$. The relative permittivity ϵ_r and relative permeability μ_r are close to unity for air. Their deviation from unity is not in itself enough to give scattering: what is important is that there are local *fluctuations* in the values of these quantities. From Maxwell's equations

$$\nabla \cdot \mathbf{D} = 0, \qquad \nabla \times \mathbf{H} = \frac{\partial \mathbf{D}}{\partial t},$$

$$\nabla \cdot \mathbf{B} = 0, \qquad \nabla \times \mathbf{E} = -\frac{\partial \mathbf{B}}{\partial t},$$

recognizing that $\mathbf{D} - \epsilon_0 \mathbf{E} = \mathbf{P}$ is the electric polarization, it follows that

$$\nabla \times (\mathbf{D} - \epsilon_0 \mathbf{E}) = \nabla \times \mathbf{D} + \epsilon_0 \dot{\mathbf{B}},$$

---

[3]The sums in Equation (11.10) are sums of geometric series.

[4]More precisely, the relative spectral intensities differ from those in white light, because the intensity of the blue-violet end of the spectrum relative to that of the red-orange end is enhanced.

[5]This is not strictly true; there are free charges in the ionosphere—but they are not responsible for the scattering that results in the blue of the sky.

so that

$$\nabla \times (\nabla \times (\mathbf{D} - \epsilon_0 \mathbf{E})) = \nabla \times (\nabla \times \mathbf{D}) + \epsilon_0 \frac{\partial}{\partial t}(\nabla \times \mathbf{B}).$$

Thus

$$\nabla \times (\nabla \times \mathbf{P}) = \nabla \times (\nabla \times \mathbf{D}) + \epsilon_0 \frac{\partial}{\partial t}(\nabla \times \mathbf{B}). \tag{11.12}$$

Likewise, using $\mathbf{B} - \mu_0 \mathbf{H} = \mu_0 \mathbf{M}$, where $\mathbf{M}$ is the magnetization, we have

$$\mu_0 \frac{\partial}{\partial t} \nabla \times \mathbf{M} = \frac{\partial}{\partial t} \left[\nabla \times \mathbf{B} - \mu_0 \frac{\partial \mathbf{D}}{\partial t} \right]. \tag{11.13}$$

Then ϵ_0 times Equation (11.13) and subtraction of Equation (11.12) gives

$$\epsilon_0 \mu_0 \frac{\partial}{\partial t} \nabla \times \mathbf{M} - \nabla \times (\nabla \times \mathbf{P}) = -\epsilon_0 \mu_0 \frac{\partial^2}{\partial t^2} \mathbf{D} - \nabla \times (\nabla \times \mathbf{D})$$

$$= \nabla^2 \mathbf{D} - \frac{1}{c^2} \frac{\partial^2}{\partial t^2} \mathbf{D}. \tag{11.14}$$

For each frequency $\omega = kc$ (i.e., with time dependence $e^{-i\omega t}$) we have then

$$(\nabla^2 + k^2)\mathbf{D}(\mathbf{x}) = \mathbf{s}(\mathbf{x}), \tag{11.15}$$

where

$$\mathbf{s} = -\frac{i\omega}{c^2} \nabla \times \mathbf{M} - \nabla \times (\nabla \times \mathbf{P}), \tag{11.16}$$

which is *small*. But it is also for the moment unknown. However, if we did know this quantity, we could use the method of Green's functions to derive

$$\mathbf{D}(\mathbf{x}) = \mathbf{D}_0(\mathbf{x}) - \frac{1}{4\pi} \int d^3 x' \frac{e^{ik|\mathbf{x}-\mathbf{x}'|}}{|\mathbf{x}-\mathbf{x}'|} \mathbf{s}(\mathbf{x}'), \tag{11.17}$$

where $\mathbf{D}_0$ is any solution to the homogeneous equation

$$(\nabla^2 + k^2)\mathbf{D}_0 = 0.$$

(Note that we have as usual imposed outgoing boundary conditions.) So if $\mathbf{D}_0$ represents the incident (plane-wave) radiation on some region where $\epsilon \neq \epsilon_0$, $\mu \neq \mu_0$, the integral represents the resultant scattered radiation from that region.

In the far zone, where $|\mathbf{x} - \mathbf{x}'| \to r$, we have

$$\mathbf{D} \to \mathbf{D}_0 + \frac{e^{ikr}}{r} \boldsymbol{\mathcal{A}}, \tag{11.18}$$

or, writing $\mathbf{x} = r\mathbf{n}$,

$$\boldsymbol{\mathcal{A}}(\mathbf{x}) = -\frac{1}{4\pi} \int d^3 x' \, e^{-ik\mathbf{n}\cdot\mathbf{x}'} \mathbf{s}(\mathbf{x}'). \tag{11.19}$$

In fact the only dependence on $\mathbf{x}$ is through its direction $\mathbf{n}$. We then obtain for the differential cross-section for scattering of radiation with polarization $\boldsymbol{\epsilon}$ from this region

$$\frac{d\sigma}{d\Omega} = \frac{|\boldsymbol{\epsilon}^* \cdot \boldsymbol{\mathcal{A}}|^2}{|\mathbf{D}_0|^2}. \tag{11.20}$$

To proceed further we need to be able to say something more about $\mathbf{s(x)}$. As remarked it is small, and is expressed in terms of the polarization $\mathbf{P} = \mathbf{D} - \epsilon_0\mathbf{E}$ and the magnetization $\mathbf{M} = \frac{\mathbf{B} - \mu_0\mathbf{H}}{\mu_0}$ above. So, on integration by parts,

$$\boldsymbol{\mathcal{A}} = -\frac{1}{4\pi} \int d^3x' \, e^{-ik\mathbf{n}\cdot\mathbf{x}'} \mathbf{s}(\mathbf{x}')$$

$$= \frac{1}{4\pi} \int d^3x' \, e^{-ik\mathbf{n}\cdot\mathbf{x}'} \left[\frac{i}{c^2}\omega\boldsymbol{\nabla} \times \mathbf{M} + \boldsymbol{\nabla} \times (\boldsymbol{\nabla} \times \mathbf{P}) \right]'$$

$$= -\frac{k^2}{4\pi} \int d^3x' \, e^{-ik\mathbf{n}\cdot\mathbf{x}'} [\mathbf{n} \times (\mathbf{n} \times \mathbf{P}') + \mathbf{n} \times \mathbf{M}'/c]. \tag{11.21}$$

We now need to be able to say something about the (induced) polarization and magnetization $\mathbf{P}$ and $\mathbf{M}$.

11.3.1 The Born Approximation

The induced polarization and magnetization will depend in detail on the medium—in this case air—and on the fields imposed. We don't know the fields, so need to know how to make some suitable approximation. We *do* know that

$$\mathbf{P} = \mathbf{D} - \epsilon_0\mathbf{E} = \epsilon_0(\epsilon_r - 1)\mathbf{E} = \epsilon_0\delta\epsilon_r\mathbf{E},$$

where $\delta\epsilon_r = \epsilon_r - 1$; and similarly

$$\mathbf{M} = (\mu_r - 1)\mathbf{H} = \frac{\delta\mu_r}{\mu}\mathbf{B},$$

where $\delta\mu_r = \mu_r - 1$. Because the scattered wave has an amplitude that is small compared with the incident wave, we may to first order in the small quantities $\delta\epsilon_r$ and $\delta\mu_r$ replace the fields $\mathbf{E}$ and $\mathbf{B}$ by the contributions to them from the incident radiation, ignoring the small contribution from the scattered radiation itself. So to first order we may write

$$\mathbf{P} \approx \delta\epsilon_r\mathbf{D}_0, \tag{11.22}$$

$$\mathbf{M} \approx \delta\mu_r c\mathbf{n}_0 \times \mathbf{D}_0, \tag{11.23}$$

where we have also used $\mathbf{B}_0 = \mathbf{n}_0 \times \mathbf{E}_0/c$ since the incident wave is taken to be a plane wave in the direction $\mathbf{n}_0$. This replacement of the full field by the contribution of the incident field alone, dropping the contribution of the scattered component, is the *Born approximation*.[6]

---

[6] In fact this is the first of a systematic system of approximations, analogous to the methods introduced by Newton to determine the perturbations induced on the Keplerian orbits of the planets by the gravitational forces of each on the others. In the context of electromagnetism, it was used by Lord Rayleigh (John William Strutt, 1814–1919) in 1881 (*Phil. Mag.* **S.5**, pp 81–101). It has wide application in the quantum theory of scattering, and it was in this context that it was introduced by Max Born (1882–1970).

Using it, we have for the Born approximation to the amplitude

$$\boldsymbol{A}_{\text{Born}} = -\frac{k^2}{4\pi} \int d^3x' \, e^{-ik\mathbf{n}\cdot\mathbf{x}'} [\mathbf{n} \times (\mathbf{n} \times \mathbf{D}_0')\delta\epsilon_r' + \mathbf{n} \times (\mathbf{n}_0 \times \mathbf{D}_0')\delta\mu_r'].$$

But if the incident wave is plane polarized with polarization vector $\boldsymbol{\epsilon}_0$ and is propagating in the direction $\mathbf{n}_0$, we have

$$\mathbf{D}_0' = \mathbf{D}_0(\mathbf{x}') = D_0\boldsymbol{\epsilon}_0 e^{ik\mathbf{n}_0\cdot\mathbf{x}'},$$

and so obtain

$$\boldsymbol{A}_{\text{Born}} = -\frac{k^2}{4\pi} D_0 \int d^3x' e^{i\mathbf{q}\cdot\mathbf{x}'} [\mathbf{n} \times (\mathbf{n} \times \boldsymbol{\epsilon}_0)\delta\epsilon_r' + \mathbf{n} \times (\mathbf{n}_0 \times \boldsymbol{\epsilon}_0)\delta\mu_r'], \quad (11.24)$$

from which it follows that

$$\left(\frac{d\sigma}{d\Omega}\right)_{\text{Born}} = \left(\frac{k^2}{4\pi}\right)^2 \left| \int d^3x \, e^{i\mathbf{q}\cdot\mathbf{x}}\boldsymbol{\epsilon}^* \cdot [\mathbf{n} \times (\mathbf{n} \times \boldsymbol{\epsilon}_0)\delta\epsilon_r + \mathbf{n} \times (\mathbf{n}_0 \times \boldsymbol{\epsilon}_0)\delta\mu_r] \right|^2$$

$$= \left(\frac{k^2}{4\pi}\right)^2 \left| \int d^3x \, e^{i\mathbf{q}\cdot\mathbf{x}}[\boldsymbol{\epsilon}^* \cdot \boldsymbol{\epsilon}_0\delta\epsilon_r + (\mathbf{n} \times \boldsymbol{\epsilon}^*) \cdot (\mathbf{n}_0 \times \boldsymbol{\epsilon}_0)\delta\mu_r] \right|^2.$$
$$(11.25)$$

A useful check with previous results is to apply this Born approximation to the scattering from a small dielectric sphere of radius a. This gives

$$\left(\frac{d\sigma}{d\Omega}\right)_{\text{Born}} = \left(\frac{k^2}{4\pi}\right)^2 \left| \int_{\text{sphere}} d^3x \, e^{i\mathbf{q}\cdot\mathbf{x}}\boldsymbol{\epsilon}^* \cdot \boldsymbol{\epsilon}_0\delta\epsilon_r \right|^2$$

$$= \left(\frac{k^2}{r4\pi}\right)^2 |\boldsymbol{\epsilon}^* \cdot \boldsymbol{\epsilon}_0|^2 (\delta\epsilon_r)^2 \left| \int_0^a r^2 \, dr \int_{-1}^{+1} d(\cos\theta) \int_0^{2\pi} d\phi \, e^{iqr\cos\theta} \right|^2$$

$$= k^4 |\boldsymbol{\epsilon}^* \cdot \boldsymbol{\epsilon}_0|^2 (\delta\epsilon_r)^2 \left(\frac{\sin qa - qa\cos qa}{q^3} \right)^2$$

$$\to k^4 a^6 |\boldsymbol{\epsilon}^* \cdot \boldsymbol{\epsilon}_0|^2 \left(\frac{\delta\epsilon_r}{3} \right)^2 \quad \text{as } qa \to 0. \quad (11.26)$$

Our previous result (see Eq. (11.8)) for the long-wavelength scattering from a small dielectric scatterer was

$$\left(\frac{k^2}{4\pi\epsilon_0}\right)^2 \frac{1}{E_0^2} |\boldsymbol{\epsilon}^* \cdot \mathbf{p}|^2.$$

For a sphere with relative permittivity ϵ_r, the induced dipole moment $\mathbf{p}$ in a static electric field may be determined to be

$$\mathbf{p} = 4\pi\epsilon_0 a^3 \left(\frac{\epsilon_r - 1}{\epsilon_r + 2} \right) \mathbf{E}_0 \approx 4\pi\epsilon_0 a^3 \left(\frac{\delta\epsilon_r}{3} \right) \mathbf{E}_0,$$

so that the two calculations agree when $\delta\epsilon_r$ is small.

11.3.2   Rayleigh's Explanation for the Blue Sky

The quantity $\delta\epsilon_r$ that appears in the considerations above is the *electric suscep-tibility* $\chi_e = \epsilon_r - 1$ of a medium. Since at optical frequencies $\delta\mu_r$ is extremely small, we may relate χ_e to the refractive index n by

$$n^2 = \epsilon_r\mu_r \approx \epsilon_r = \chi_e + 1.$$

This is not yet enough to allow us to use the previous discussion to give an ex-planation for the scattering of sunlight in the atmosphere if the air is considered as a homogeneous medium. As already indicated, it is best explained through density fluctuations, but we give first another, somewhat simpler explanation due to Rayleigh. The basis of the discussion is to recognize that the air is not homogeneous, because it is constituted from molecules, and what we should do is to consider the scattering from the random distribution of molecules, each of which may be thought of as an electrically polarizable dipole. This model gives for the electric susceptibility

$$\chi_e(\mathbf{x}) = \frac{1}{\epsilon_0}\sum_j \gamma_{\mathrm{mol}}\delta^{(3)}(\mathbf{x} - \mathbf{x}_j), \qquad (11.27)$$

where the *molecular polarizability* γ_{mol} is defined so that each molecule has an induced dipole moment

$$\mathbf{p}_j = \gamma_{\mathrm{mol}}\mathbf{E}(\mathbf{x}_j)$$

determined by the field $\mathbf{E}(\mathbf{x}_j)$ it experiences. This field is of course affected by the presence of other (polarized) molecules, and we will return to its determi-nation below. But what we can already do is to use previous results and the Born approximation to obtain[7]

$$\frac{d\sigma}{d\Omega} = \left(\frac{k^2}{4\pi}\right)^2 \left| \int d^3x\, e^{i\mathbf{q}\cdot\mathbf{x}}\boldsymbol{\epsilon}^* \cdot \boldsymbol{\epsilon}_0\frac{1}{\epsilon_0}\sum_j \gamma_{\mathrm{mol}}\delta^{(3)}(\mathbf{x} - \mathbf{x}_j) \right|^2$$

$$= \left(\frac{k^2}{4\pi\epsilon_0}\right)^2 |\boldsymbol{\epsilon}^* \cdot \boldsymbol{\epsilon}_0|^2\gamma_{\mathrm{mol}}^2 \left| \sum_j e^{i\mathbf{q}\cdot\mathbf{x}_j} \right|^2$$

$$= \left(\frac{k^2}{4\pi\epsilon_0}\right)^2 |\boldsymbol{\epsilon}^* \cdot \boldsymbol{\epsilon}_0|^2\gamma_{\mathrm{mol}}^2 N, \qquad (11.28)$$

since the molecules are randomly distributed, and the resultant scattering is incoherent. N is the number of molecules, and for simplicity they have all been taken to have the same polarizability. Thus the differential scattering coss-section *per molecule* is

$$k^4|\boldsymbol{\epsilon}^* \cdot \boldsymbol{\epsilon}_0|^2 \left(\frac{\gamma_{\mathrm{mol}}}{4\pi\epsilon_0}\right)^2.$$

---

[7]Note that here and in the equations that follow, the symbol ϵ_0 is the permittivity of the vacuum (Eq. (3.7)), whilst $\boldsymbol{\epsilon}_0$ and $\boldsymbol{\epsilon}^*$ are polarization vectors.

The next step is to relate the molecular polarizability to the bulk polar-izablity, and hence to the susceptibility and the refractive index. Evidently the bulk polarization of a volume V of the medium containing N molecules is $\mathbf{P}V = N\langle\mathbf{p}_{\text{mol}}\rangle$, that is, N times the average molecular polarization. ($\mathbf{P}$ is af-ter all defined to be the average induced dipole moment per unit volume.) The polarization of an individual molecule is $\mathbf{p}_{\text{mol}} = \gamma_{\text{mol}}\mathbf{E}_{\text{mol}}$, so now the problem is to determine the *molecular field* $\mathbf{E}_{\text{mol}}$, or rather its average.

Consider a slab of a dielectric medium placed in a uniform electric field $\mathbf{E}_{\text{ext}}$, for example, a slab of dielectric between the plates of a parallel plate capacitor. There will be a surface charge on the plates of the capacitor, σ, and the field is $\mathbf{E}_{\text{ext}} = \frac{1}{\epsilon_0}\sigma\mathbf{n}$, where $\mathbf{n}$ is the unit vector normal to the plates. The dielectric slab is polarized by this field; each molecule becomes an electric dipole, and there results an induced charge on the surface of the dielectric opposite in sign to the charge on the adjacent plate of the capacitor. The density of this induced charge is $\sigma_p = -P$, as follows from the definition of P as the magnitude of the polarization of the dielectric. These induced charges on the faces of the dielectric slab set up an electric field $\mathbf{E}_{\text{dep}} = \frac{1}{\epsilon_0}\sigma_p\mathbf{n} = -\frac{1}{\epsilon_0}\mathbf{P}$ that acts in the sense to depolarize the medium. The *macroscopic* field inside the slab is $\mathbf{E}_{\text{mac}} = \mathbf{E}_{\text{ext}} + \mathbf{E}_{\text{dep}} = \mathbf{E}_{\text{ext}} - \frac{1}{\epsilon_0}\mathbf{P}$, or

$$\epsilon_0\mathbf{E}_{\text{ext}} = \epsilon_0\mathbf{E}_{\text{mac}} + \mathbf{P}.$$

Now imagine a small spherical cavity excavated from the interior of the dielectric (Fig. 11.1).

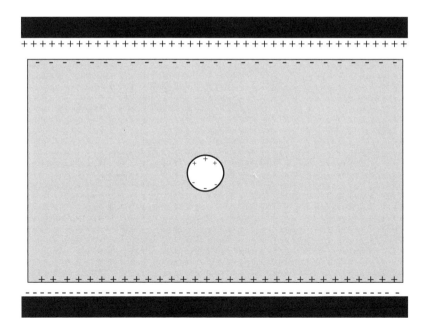

Figure 11.1. A cavity in a dielectric slab between oppositely charged plates.

There would be induced charges on its surface, which would lead to a contribution to the electric field at the centre of the cavity of

$$\mathbf{E}_{\text{surf}} = \frac{1}{3}\frac{1}{\epsilon_0}\mathbf{P}.$$

The factor of $\frac{1}{3}$ comes about from the fact that the cavity is a *sphere*, and the sign is consistent with the fact that the induced charges are such as to give rise to a field in the same sense as the original external field. So at the center of the cavity the field is

$$\begin{aligned}
\mathbf{E}_{\text{cav}} &= \mathbf{E}_{\text{ext}} + \mathbf{E}_{\text{dep}} + \mathbf{E}_{\text{surf}} \\
&= \left(\mathbf{E}_{\text{mac}} + \frac{1}{\epsilon_0}\mathbf{P}\right) + \left(-\frac{1}{\epsilon_0}\mathbf{P}\right) + \left(\frac{1}{3\epsilon_0}\mathbf{P}\right) \\
&= \mathbf{E}_{\text{mac}} + \frac{1}{3\epsilon_0}\mathbf{P}.
\end{aligned}$$

Now replace the material excavated from the cavity. A molecule at the center of the sphere experiences a field

$$\mathbf{E}_{\text{mol}} = \mathbf{E}_{\text{cav}} + \mathbf{E}_{\text{near}},$$

where $\mathbf{E}_{\text{near}}$ is the field due to the other (polarized) molecules inside the sphere. This is again difficult to determine in the general case, but for certain simple cases, it can be shown to vanish. In particular, for a random distribution of molecules, or for a simple cubic crystal (as was shown by Lorentz), it vanishes, and we are justified on the former ground for taking it to be zero in the case of the air in which we are interested. So we have

$$\langle\mathbf{E}_{\text{mol}}\rangle = \langle\mathbf{E}_{\text{cav}}\rangle = \mathbf{E}_{\text{mac}} + \frac{1}{3\epsilon_0}\mathbf{P}.$$

But we also have

$$\langle\mathbf{P}_{\text{mol}}\rangle = \gamma_{\text{mol}}\langle\mathbf{E}_{\text{mol}}\rangle = \gamma_{\text{mol}}\left(\mathbf{E}_{\text{mac}} + \frac{1}{3\epsilon_0}\mathbf{P}\right),$$

and since $\mathbf{P} = \frac{N}{V}\langle\mathbf{P}_{\text{mol}}\rangle$, we obtain

$$\mathbf{P} = \rho_N\gamma_{\text{mol}}\left(\mathbf{E}_{\text{mac}} + \frac{1}{3\epsilon_0}\mathbf{P}\right),$$

$\rho_N = N/V$ being the (average) number of molecules in a unit volume. On the other hand, we also have $\mathbf{D} = \epsilon\mathbf{E}$, or better put $\mathbf{D} = \epsilon\mathbf{E}_{\text{mac}}$, and since $\mathbf{D}$ is constant throughout the region between the plates of the capacitor (there are no free charges), we have

$$\mathbf{D} = \epsilon_0\mathbf{E}_{\text{ext}} = \epsilon_0\mathbf{E}_{\text{mac}} + \mathbf{P}.$$

So
$$\mathbf{P} = \mathbf{D} - \epsilon_0 \mathbf{E}_{\mathrm{mac}} = (\epsilon_r - 1)\epsilon_0 \mathbf{E}_{\mathrm{mac}}.$$

This means that we have (at last!) derived
$$(\epsilon_r - 1)\epsilon_0 = \rho_N \gamma_{\mathrm{mol}} \left[1 + \frac{1}{3}(\epsilon_r - 1) \right]$$

or
$$\gamma_{\mathrm{mol}} = \frac{3\epsilon_0}{\rho_N} \left(\frac{\epsilon_r - 1}{\epsilon_r + 2} \right), \tag{11.29}$$

a result due to Clausius (1850) and Mossotti (1879). Using the previously given relation between ϵ_r and the refractive index, one obtains

$$\frac{\gamma_{\mathrm{mol}}}{4\pi\epsilon_0} = \frac{3}{4\pi\rho_N} \frac{n^2 - 1}{n^2 + 2}, \tag{11.30}$$

known as the *Lorentz-Lorenz equation* (1880). This can now be used to express the differential scattering of light from the air in the sky as

$$\left(\frac{d\sigma}{d\Omega} \right)_{\mathrm{per\ mol}} = k^4 |\boldsymbol{\epsilon}^* \cdot \boldsymbol{\epsilon}_0|^2 \left(\frac{3}{4\pi\rho_N} \frac{n^2 - 1}{n^2 + 2} \right)^2$$

$$\approx k^4 |\boldsymbol{\epsilon}^* \cdot \boldsymbol{\epsilon}_0|^2 \left(\frac{1}{2\pi\rho_N} \right)^2 (n - 1)^2, \tag{11.31}$$

where at the last step we have used $(n - 1) \ll 1$.

So far in this long calculation we have considered the scattering of incident sunlight which has been taken to be polarized with polarization vector $\boldsymbol{\epsilon}_0$. Since sunlight is unpolarized, we should really take an *average* over the two independent directions normal to $\mathbf{n}_0$. This then gives for the differential cross-section for scattering leading to a polarization *parallel* to the *scattering plane* defined by the vectors $\mathbf{n}$ and $\mathbf{n}_0$

$$\left(\frac{d\sigma}{d\Omega} \right)_{\parallel} = \frac{1}{2} k^4 \left(\frac{1}{2\pi\rho_N} \right)^2 (n - 1)^2 \cos^2 \theta, \tag{11.32}$$

and for scattering leading to a polarization *perpendicular* to the scattering plane

$$\left(\frac{d\sigma}{d\Omega} \right)_{\perp} = \frac{1}{2} k^4 \left(\frac{1}{2\pi\rho_N} \right)^2 (n - 1)^2. \tag{11.33}$$

The scattered light is thus polarized on average, with a polarization given by the difference between the perpendicular and parallel cross-sections divided by their sum, that is,

$$\Pi(\theta) = \frac{1 - \cos^2 \theta}{1 + \cos^2 \theta} = \frac{\sin^2 \theta}{1 + \cos^2 \theta}. \tag{11.34}$$

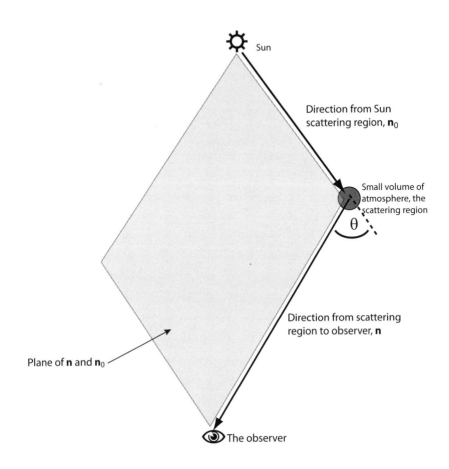

Figure 11.2. Sunlight scattered from the atmosphere.

Here, as usual, θ is the angle of scattering, that is, the angle between the incident direction $\mathbf{n}_0$ and the scattered direction $\mathbf{n}$.

The *total* differential cross-section is the sum of the two possible final polarizations and so is

$$\left(\frac{d\sigma}{d\Omega}\right)_T = \frac{1}{2}k^4 \left(\frac{1}{2\pi\rho_N}\right)^2 (n-1)^2(1+\cos^2\theta). \tag{11.35}$$

On integrating this over angles, one obtains for the total cross-section per molecule

$$\sigma = \frac{2}{3\pi}\frac{k^4}{\rho_N^2}(n-1)^2. \tag{11.36}$$

This is the power scattered out of the incident sunlight per molecule per unit incident flux. Then if I is the incident intensity, in traversing a thickness dx of the atmosphere, the intensity is reduced by $dI = -I\rho_N\sigma\,dx$ so that after traversing a thickness x the intensity is given by

$$I = I_0 e^{-\alpha x},$$

where α, the *attenuation coefficient*, is given by

$$\alpha = \rho_N\sigma = \frac{2}{3\pi}\frac{k^4}{\rho_N}(n-1)^2, \tag{11.37}$$

with a corresponding expression for its inverse Λ, the *attenuation length*. This expression for the attenuation length is due to Rayleigh. Note especially that the attenuation length is proportional to ρ_N, the number of molecules per unit volume. If there were no atomicity, there would be no attenuation, and the sky would not be blue! Note also the k^4 factor, which leads to very substantial differences in Λ across the spectrum of visible wavelengths. There is also a wavelength dependence in the refractive index n.

It is of course the scattered light which, because of the k^4 factor, is perceived as blue. The variation in polarization from zenith to horizon is not apparent to our eyes (unless viewed with polarizing sunglasses), but is used, for example, by bees to navigate.

11.4   Critical Opalescence

The discussion above is not appropriate to dense media, where the scattering is primarily caused by fluctuations in density. To deal with this situation, consider a division of the volume V from which the scattering occurs into small cells of volume v. The cells, although small, with linear dimension much less than a wavelength, are to be large enough each to contain many molecules. The number of molecules in each cell will *on average* be $\rho_N v$, where ρ_N is, as before, the (average) number of molecules per unit volume. The fluctuations in density will result in the number of molecules in the jth cell being different from the average number $\rho_N v$; let this difference be Δ_j, that is, there are $\rho_N v + \Delta_j$

molecules in the jth cell. The small difference in density of the jth cell produces a corresponding difference

$$\delta\epsilon_{rj} = \frac{\partial\epsilon_r}{\partial\rho_N}\frac{\Delta_j}{v}$$

in the relative permittivity of the jth cell. But by differentiating the logarithm of the Clausius-Mossotti equation,

$$\frac{\epsilon_r - 1}{\epsilon_r + 2} = \frac{4\pi\rho_N}{3}\frac{\gamma_{\text{mol}}}{4\pi\epsilon_0},$$

one obtains

$$\frac{\partial\epsilon_r}{\partial\rho_N} = \frac{(\epsilon_r - 1)(\epsilon_r + 2)}{3\rho_N}.$$

Thus we have

$$\delta\epsilon_{rj} = \frac{(\epsilon_r - 1)(\epsilon_r + 2)}{3\rho_N}\frac{\Delta_j}{v}.$$

Recall that (setting $\delta\mu_r = 0$, as is appropriate at optical frequencies) for a single scatterer

$$\frac{d\sigma}{d\Omega} = \left(\frac{k^2}{4\pi}\right)^2 \left| \int d^3x\, e^{i\mathbf{q}\cdot\mathbf{x}}\boldsymbol{\epsilon}^* \cdot \boldsymbol{\epsilon}_0\, \delta\epsilon_r \right|^2.$$

Treating each cell as a single scatterer, and including the contributions from all V/v cells in the volume V, which contains a total of $\rho_N V$ molecules, we have

$$\frac{d\sigma}{d\Omega} = \left(\frac{k^2}{4\pi}\right)^2 \left| \sum_j \int_{v_j} d^3x\, e^{i\mathbf{q}\cdot\mathbf{x}_j}\boldsymbol{\epsilon}^* \cdot \boldsymbol{\epsilon}_0\, \delta\epsilon_r \right|^2$$

$$= \left(\frac{k^2}{4\pi}\right)^2 \left| \sum_j v\,\boldsymbol{\epsilon}^* \cdot \boldsymbol{\epsilon}_0\, \delta\epsilon_{rj} \right|^2$$

$$= \left(\frac{k^2}{4\pi}\right)^2 |\boldsymbol{\epsilon}^* \cdot \boldsymbol{\epsilon}_0|^2 \left| \sum_j v\frac{(\epsilon_r - 1)(\epsilon_r + 2)}{3\rho_N v}\Delta_j \right|^2$$

$$= \left(\frac{k^2}{4\pi}\right)^2 |\boldsymbol{\epsilon}^* \cdot \boldsymbol{\epsilon}_0|^2 \left[\frac{(\epsilon_r - 1)(\epsilon_r + 2)}{3\rho_N}\right]^2 \left| \sum_j \Delta_j \right|^2. \tag{11.38}$$

In going from the first to the second line, we have assumed that the correlation between fluctuations from one cell to the next extends only over distances small compared with the wavelength, so that the exponential can be set to unity. Summing over the initial polarizations $\boldsymbol{\epsilon}_0$, averaging over the final polarizations $\boldsymbol{\epsilon}$, and integrating over $d\Omega$, we obtain for the total cross-section for scattering from the $\rho_N V$ molecules in the volume V:

$$\sigma = \left(\frac{k^2}{4\pi}\right)^2 \left(\frac{8\pi}{3}\right) \left[\frac{(\epsilon_r - 1)(\epsilon_r + 2)}{3\rho_N}\right]^2 \left| \sum_j \Delta_j \right|^2. \tag{11.39}$$

The cross-section *per molecule* is then obtained by dividing by $\rho_N V$, and the attenuation coefficient α, which is ρ_N times the cross-section per molecule, is

$$\alpha = \left(\frac{k^2}{4\pi}\right)^2 \left(\frac{8\pi}{3}\right) \left[\frac{(\epsilon_r - 1)(\epsilon_r + 2)}{3\rho_N}\right]^2 \frac{|\sum_j \Delta_j|^2}{V}.$$

Now the quantity $|\sum_j \Delta_j|^2 \equiv \Delta_V^2$ is the square of the sum of the fluctuation in the number of molecules in each cell taken over all the cells in the volume V, that is, it is the square of the fluctuation in the number of molecules in the volume V from the average number $\rho_N V$ of molecules in that volume. That is in turn given by statistical mechanics arguments as[8]

$$\frac{\Delta_V^2}{\rho_N V} = \rho_N k_B T \beta_T,$$

where T is the temperature, k_B is the Boltzmann constant, and β_T is the isothermal compressibility, $\beta_T = -\frac{1}{V}\left(\frac{\partial V}{\partial P}\right)_T$. From this it follows that

$$\alpha = \left(\frac{k^2}{4\pi}\right)^2 \left(\frac{8\pi}{3}\right) \left[\frac{(\epsilon_r - 1)(\epsilon_r + 2)}{3\rho_N}\right]^2 \rho_N^2 k_B T \beta_T$$

$$= k^4 \frac{2}{3\pi\rho_N} \left[\frac{(\epsilon_r - 1)(\epsilon_r + 2)}{6}\right]^2 (\rho_N k_B T \beta_T). \tag{11.40}$$

This is the Einstein-Smoluchowski equation (Smoluchowski obtained an incomplete version in 1904, which was completed by Einstein in 1910). The factor $|(\epsilon_r - 1)(\epsilon_r + 2)/6|$ is well approximated by $n - 1$ when $|\epsilon_r - 1| \ll 1$, and for an ideal gas, $\rho_N k_B T \beta_T = 1$, so that we recover the earlier result

$$\alpha = k^4 \frac{2}{3\pi\rho_N}(n - 1)^2$$

of Rayleigh. But we also learn that, as the temperature approaches a critical temperature at which β_T becomes infinite, the attenuation length diverges, and this leads to the phenomenon of *critical opalescence*. This isn't quite the end of the story, since the approximations we have made break down near the critical point. For as the critical point is approached, the correlation length for density fluctuations diverges, and so exceeds the wavelength; and we assumed no correlation from one cell to the other. The more detailed consideration needed to complete the argument was given in 1914 by Ornstein and Zernicke; but that is outside the scope of this book.

---

[8]The *Einstein equation* to be found (in different notation) in *Ann. Phys.* **33**, 1275 (1910).

Chapter 12

Dispersion

As we saw in Section 10.1, the spreading out of the different colors in the spectrum as observed by Newton, their *dispersion* into an oblong rather than the circular form he had expected, is a consequence of the frequency dependence of the speed of propagation of light in the glass of his prism. The wave-fronts of a plane wave $e^{i(\mathbf{k}\cdot\mathbf{x}-\omega t)}$ propogate with the *phase velocity* $\mathbf{v}$ or speed $v = \frac{\omega}{k}$, and this depends on ω in a dispersive medium. The relationship between frequency ω and wave number k is called a *dispersion relation*, and when it is not linear, the waves suffer dispersion. For electromagnetic waves in vacuum where $\omega = ck$ there is no dispersion. But in a medium with

$$\left(\frac{\omega}{k}\right)^2 = v^2 = \frac{c^2}{\epsilon_r \mu_r},$$

the frequency dependence of $\epsilon_r \mu_r = n^2$ leads to dispersion.

Thus dispersion and associated phenomena are consequences of the electric polarizability of the medium (at optical frequencies we may ignore magnetic effects). So to try better to understand their microscopic origin we will give a classical model for the polarization $\mathbf{P}$. This model, described by Lorentz[1] in the 1890s, predates quantum mechanics, but was influential on the work of Einstein and Planck, and indeed on Heisenberg, in establishing the corresponding quantum-mechanical understanding of the theory. Its success derives from the fact that it captures the essential physics of the polarzability of matter. Although the medium is electrically neutral on the macroscopic scale, at the atomic level it has electrically charged constituents. This is already apparent from the nature of electrolysis (studied by Faraday), and the later discovery of the electron.

12.1 The Oscillator Model

Following Lorentz, we will suppose that electrons are bound in the molecules of the medium by some sort of elastic force. Simplifying further to consider only

---

[1] Hendrik Antoon Lorentz (1853–1928).

their motion in the x-direction, the equation of motion (Newton's equation!) for such an electron will be

$$m\frac{d^2x}{dx^2} = -m\omega_0^2 x.$$

The electron oscillates with frequency ω_0. If now the molecule is subject to a monochromatic oscillating electric field, also in the x-direction, with frequency ω, so that $E(t) = E(\omega)e^{-i\omega t}$, the equation of motion is modified to become that for a damped driven harmonic oscillator[2]

$$m\frac{d^2x}{dx^2} = -2m\alpha\frac{dx}{dt} - m\omega_0^2 x - eE(t),$$

with solution for $x(t) = x(\omega)e^{-i\omega t}$

$$x(\omega) = \frac{eE(\omega)}{m(\omega^2 + 2i\alpha\omega - \omega_0^2)}. \tag{12.1}$$

(α is a phenomenological damping coefficient.) This means that the molecule responds to the oscillating electric field with an oscillating dipole moment $p(\omega)e^{-i\omega t}$ with

$$p(\omega) = -ex(\omega) = -\frac{e^2}{m}\frac{1}{2\omega_r}\left(\frac{1}{\omega - \omega_r + i\alpha} - \frac{1}{\omega + \omega_r + i\alpha}\right)E(\omega).$$

Here ω_r is the *resonance frequency* shifted from ω_0 due to the damping, $\omega_r = \sqrt{\omega_0^2 - \alpha^2}$. The contribution to the molecular polarizability is defined again by $p = \gamma_{\text{mol}}E$ with E the electric field strength experienced by the molecule.

The resonance frequency might be supposed to be related to the absorption or emission frequency observed in the line spectrum of the medium. But there is not just one such line in the spectrum, and indeed very many more lines are observed than there are electrons in the molecule. A mechanical model with the electrons having a number of modes of oscillation might be suggested. But none of this is necessary, since a quantum-mechanical calculation, while giving a result that is essentially similar to what we have just obtained, provides a modification that replaces the formula (11.29) for γ_{mol} with

$$\gamma_{\text{mol}}(\omega) = -\sum_r \frac{e^2}{2m\omega_r}f_r\left(\frac{1}{\omega - \omega_r + i\alpha_r} - \frac{1}{\omega + \omega_r + i\alpha_r}\right).$$

The *oscillator strengths* f_r can be calculated from the quantum theoretical formulation, and satisfy various *sum rules*.[3] Once again, we ought to recognize that

---

[2] Charge of electron $= -e$.

[3] Of these the most celebrated is the Thomas-Reiche-Kuhn sum rule

$$\sum_j f_j = Z.$$

Here Z is the number of electrons in the molecule, and the sum is over all the quantum states j that can be reached by a dipole transition from the ground state.

the (average) electric field experienced by the molecules is not the externally imposed field E_{mac} but rather it is $\langle E_{\text{mol}} \rangle = E_{\text{mac}} + \frac{1}{3\epsilon_0}P$, giving the previously derived Clausius-Mossotti/Lorentz-Lorenz relations between γ_{mol}, the relative permittivity ϵ_r, and the refractive index n:

$$\gamma_{\text{mol}} = \frac{3\epsilon_0}{\rho_N}\left(\frac{\epsilon_r - 1}{\epsilon_r + 2}\right) = \frac{3\epsilon_0}{\rho_N}\left(\frac{n^2 - 1}{n^2 + 2}\right)$$

or

$$n^2 = \epsilon_r = \frac{1 + \frac{2}{3}\frac{\rho_N}{\epsilon_0}\gamma_{\text{mol}}}{1 - \frac{1}{3}\frac{\rho_N}{\epsilon_0}\gamma_{\text{mol}}}. \tag{12.2}$$

For most dielectrics (including dilute gases) the dominant effect of replacing E_{mac} by E_{mol} is a downward shift in the resonance frequencies, so that with this adjustment, the model is well represented by

$$n^2 = \epsilon_r = 1 - \frac{\rho_N}{\epsilon_0}\sum_r \frac{e^2}{2m\omega_r}f_r\left(\frac{1}{\omega - \omega_r + i\alpha_r} - \frac{1}{\omega + \omega_r + i\alpha_r}\right). \tag{12.3}$$

But in any case we can now understand the origin of the frequency dependence of the refractive index; it is traceable back to the frequency dependence of the molecular polarizability.

There are some features of this model that survive in the more complete treatment provided by quantum mechanics. First, the refractive index is complex, with a *positive* imaginary part. This implies that a traveling wave with frequency ω, having the usual dependence $\exp[i(kx - \omega t)]$, is attenuated as it progresses, because

$$k(\omega) = \frac{\omega}{c}n(\omega)$$

acquires a positive imaginary part, and with the separation of k into its real and imaginary parts, $k = k_R + ik_I$, we have

$$\exp[i(kx - \omega t)] = \exp[-k_I x]\exp[i(k_R x - \omega t)],$$

which decays exponentially as x increases. This is of course what was to be expected because the oscillators were damped. Another general property is that, except when ω is close to one of the resonances, the real part of the refractive index increases and the imaginary part is small. What this means is that the medium is effectively transparent away from the resonance frequencies. For the gases in the atmosphere, these frequencies associated with the electrons are in the ultraviolet. But there are also analogous effects caused by the motion of the atomic nuclei in molecules which occur at the much lower frequencies of the infrared, because the nuclei are much more massive than the electrons. This explains why the transparency of the atmosphere (and also of water and glass) is effectively restricted to the "window" of frequencies in the visible range. The "normal" increase of the refractive index from lower to higher frequencies accounts for the order of colors in the spectrum produced by a glass prism.

However, whenever ω increases through a resonance the real part of n first *decreases*[4] and then rises rapidly. At the same time its imaginary part, and

---

[4]This is called *anomalous* dispersion.

so also the absorption, rises and falls. Resonant absorption and anomalous dispersion go hand-in-hand. The fall of the real part of the refractive index brings it below 1, and $n_R < 1$ seems to suggest the speed of propagation v exceeds c, the speed of light. But the formula $v = \omega/k$ gives the phase velocity v_{phase}, not the group velocity, which is $v_{\text{group}} = d\omega/dk$, and this never exceeds c.

12.1.1   The High-Frequency Limit

In the limit of high frequencies, higher than any of the resonance frequencies, we have

$$\epsilon_r(\omega) = n^2 \to 1 - \frac{\rho_N}{\epsilon_0}\frac{e^2}{m}\sum_r f_r \frac{1}{\omega^2}$$

$$= 1 - \frac{\omega_P^2}{\omega^2}, \tag{12.4}$$

where ω_P is called the *plasma frequency*. Note that the limit is apporoached from below.

The electrons in a *plasma* are essentially free, which means that at all but the lowest frequencies the high-frequency limit can be used. For frequencies ω less than the plasma frequency this implies that the refractive index n becomes purely imaginary, which as we have seen implies attenuation of the radiation. Radiation at such frequencies impinging on a plasma penetrates very little and is for the most part reflected.

The *ionosphere*, the upper layer of the atmosphere, is ionized by solar radiation, and so is effectively a dilute but fully ionized plasma, and its reflection of HF (high frequency = short wave) radio frequency radiation allows transmitted signals to be "bounced" from it and so enables communication between distant points beyond line-of-sight range.

12.1.2   The Drude Model

In 1900 Paul Drude (1863–1906) proposed a model[5] for the conductivity (both electric and thermal) and optical properties of metals. The valence electrons in a metal also are essentially "free," and the theory we have outlined above can be applied also to metals, giving an effective expression for the relative permittivity

$$\epsilon_r = \epsilon_r' - \frac{\rho_f}{\epsilon_0}\frac{e^2}{m}\frac{1}{\omega^2 + 2i\alpha_f\omega},$$

in which ρ_f is now the number density of free electrons, with damping coefficient α_f, and ϵ_r' represents the contribution from all the other oscillators. At low frequencies, the contribution from the free electrons dominates, so that as ω approaches zero

$$\epsilon_r \sim \frac{\rho_f}{\epsilon_0}\frac{e^2}{m}\frac{1}{2\alpha_f\omega}.$$

---

[5] "On the Electron Theory of Metals", *Ann. Phys.* **306**, 566–613 (1900), and **308**, 369–402 (1900).

But the zero-frequency *conductivity* $\sigma(\omega = 0)$ of the metal can also be related to the damping term $2\alpha_f$ that acts on the conduction electrons. Although the conduction electrons experience an acceleration $-e\mathbf{E}/m$ they also suffer collisions. After a collision the electron's velocity may, as in diffusion, be stochastically averaged to zero. This means that with an average time between collisions equal to τ say, there results a constant drift velocity $\mathbf{v}_{\text{drift}} = -e\tau\mathbf{E}/m$. And this in turn is related to the DC conductivity $\sigma(0)$, because the current is $\mathbf{j} = -e\rho_f\mathbf{v}_{\text{drift}}$ and also $\mathbf{j} = \sigma(0)\mathbf{E}$. Putting these together yields

$$\sigma(0) = e^2\rho_f\tau.$$

The average damping force on a free electron is $2m\alpha_f\mathbf{v}_{\text{drift}}$ and this must cancel the accelerating force produced by the electric field, so that we have

$$2\alpha_f = \frac{1}{\tau} = \frac{e^2\rho_f}{m\sigma(0)},$$

and conclude that

$$\epsilon_r \sim \frac{i\sigma(0)}{\epsilon_0\omega}. \tag{12.5}$$

Moving away from zero frequency, the contribution to ϵ_r from the free electrons is

$$\frac{\rho_f e^2}{m\epsilon_0}\frac{1}{\omega}\frac{i}{2\alpha_f - i\omega}.$$

This can also be used to provide an equation for the frequency dependence of the conductivity,

$$\sigma(\omega) = \sigma(0)(1 - i\omega\tau)^{-1}.$$

Returning to the Lorentz model for the optical properties of the metal, we have

$$n^2 = \epsilon_r = 1 + \frac{\rho_f e^2}{m\epsilon_0}\frac{1}{\omega}\frac{i}{2\alpha_f - i\omega} - \frac{\rho_N e^2}{m\epsilon_0}\sum_{r\neq f} f_r\frac{1}{2\omega_r}\left(\frac{1}{\omega - \omega_r + i\alpha_r} - \frac{1}{\omega + \omega_r + i\alpha_r}\right),$$

$$\tag{12.6}$$

where we have separated out the contribution from the free electrons from the oscillators $(r \neq f)$. At low frequencies the free-electron contribution, as we have said, will dominate, and when both $\omega \ll \frac{1}{\tau}$ and $\omega \ll \frac{\sigma}{\epsilon_0}$ are satisfied (for example, ω less than about 10^{12} Hz for copper) the only relevant term is the free-electron term, and n^2 is pure imaginary. Taking the square root (using $\sqrt{i} = (1 + i)/\sqrt{2}$) shows that both the real and imaginary parts of the refractive index n are large. Electromagnetic waves of low frequencies are rapidly attenuated in the metal—their penetration is essentially restricted to within the *skin depth*. The metal is shiny—it reflects well. The metal is a conductor of electricity (and of heat). It obeys Ohm's law.

At high frequencies, the approximation of the previous subsection becomes appropriate, and the plasma approximation

$$n^2 \approx 1 - \frac{\omega_P^2}{\omega^2}$$

is again valid. This implies that for frequencies above the plasma frequency ω_P (typically in the ultraviolet) the imaginary part of n^2 is negligible, and the metal becomes *transparent*.

12.2   Dispersion Relations

We can now exploit our model for the dependence of ϵ_r on frequency ω. Let us start from

$$\mathbf{D}(\mathbf{x}, \omega) = \epsilon_0 \epsilon_r(\omega) \mathbf{E}(\mathbf{x}, \omega),$$

which implies

$$\mathbf{D}(\mathbf{x}, t) - \epsilon_0 \mathbf{E}(\mathbf{x}, t) = \frac{\epsilon_0}{2\pi} \int_{-\infty}^{\infty} [\epsilon_r(\omega) - 1] \mathbf{E}(\mathbf{x}, \omega) e^{-i\omega t} \, d\omega$$

$$= \epsilon_0 \int_{-\infty}^{\infty} G(\tau) \mathbf{E}(\mathbf{x}, t - \tau) \, d\tau, \qquad (12.7)$$

where we have introduced

$$G(\tau) = \frac{1}{2\pi} \int_{-\infty}^{\infty} [\epsilon_r(\omega) - 1] e^{-i\omega\tau} \, d\omega.$$

The last step follows from the convolution theorem of Fourier integrals, which states that under suitable conditions (here assumed), the Fourier transform of the product of two functions (in this case $[\epsilon_r(\omega) - 1]$ and $\mathbf{E}(\mathbf{x}, \omega)$) is the convolution of their Fourier transforms (so of $G(\tau)$ and $\mathbf{E}(\mathbf{x}, t)$).

Our model for the frequency dependence of ϵ_r considered what happens if the medium is made of molecules with a single oscillator electron:

$$\epsilon_r = 1 + \frac{\omega_P^2}{\omega_0^2 - \omega^2 - 2i\alpha\omega}.$$

Then we have

$$G(\tau) = \frac{\omega_P^2}{2\pi} \int_{-\infty}^{\infty} \frac{e^{-i\omega\tau}}{\omega_0^2 - \omega^2 - i\gamma\omega} \, d\omega.$$

The integral may be evaluated by the usual method of completing a contour in the complex plane; the integrand has poles at $\omega = \omega_\pm = -\frac{i\gamma}{2} \pm \omega_R$ with $\omega_R^2 = \omega_0^2 - \alpha^2$. These poles lie in the lower half of the complex ω-plane. For $\tau < 0$, the factor $e^{-i\omega\tau}$ tends to zero along a semicircle at infinity in the upper half-plane, and since there are no singularities in the upper half-plane, taking as contour of integration the familiar one composed from the real axis and the arc at infinity in the upper half-plane, we conclude that for $\tau < 0$, $G(\tau) = 0$. On the other hand for $\tau > 0$ we must close the contour in the lower half-plane to get convergence on the arc at infinity, and the contour then encloses the poles of the integrand (in the negative sense, which introduces an extra minus sign), giving with use of the residue theorem

$$
\begin{aligned}
G(\tau) &= -\frac{\omega_P^2}{2\pi} \oint \frac{e^{-i\omega\tau}}{-(\omega - \omega_+)(\omega - \omega_-)}\, d\omega \\
&= -\frac{\omega_P^2}{2\pi}(2\pi i)\left[\frac{e^{-i\omega_+\tau}}{-(\omega_+ - \omega_-)} + \frac{e^{-i\omega_-\tau}}{-(\omega_- - \omega_+)}\right] \\
&= -\frac{\omega_P^2}{2\pi}(2\pi i)e^{-i(-i\gamma\tau/2)}\left[\frac{e^{-i\omega_R\tau}}{-2\omega_R} + \frac{e^{+i\omega_R\tau}}{2\omega_R}\right] \\
&= \omega_P^2\, e^{-\gamma\tau/2}\frac{\sin\omega_R\tau}{\omega_R},
\end{aligned} \tag{12.8}
$$

for $\tau > 0$. Putting these results together, we have

$$
G(\tau) = \omega_P^2\, e^{-\gamma\tau/2}\frac{\sin\omega_R\tau}{\omega_R}\theta(\tau).
$$

The details are not so important as the presence of the factor $\theta(\tau)$, which is clearly still there even in the more sophisticated model in which there was a sum over similar terms (many oscillators, each contributing a term as above with an oscillator strength), or indeed in the full quantum-mechanical treatment (which in effect gives an explanation of the oscillator strengths).

There is a nonlocality in time in the relation between $\mathbf{D}$ and $\mathbf{E}$ by of order $\alpha^{-1} \approx 10^{-8}$ s for a typical spectral line; α is the natural linewidth. The factor $\theta(\tau)$ is essentially independent of the model, and it ensures that the response of the dielectric is *causally* related to the $\mathbf{E}$-field,

$$
\frac{1}{\epsilon_0}\mathbf{D}(\mathbf{x}, t) = \mathbf{E}(\mathbf{x}, t) + \int_0^\infty G(\tau)\mathbf{E}(\mathbf{x}, t - \tau)\, d\tau,
$$

since the lower limit of the τ-integration may be taken as 0, not $-\infty$. This then means that

$$
\epsilon_r(\omega) = 1 + \int_0^\infty G(\tau)e^{i\omega\tau}\, d\tau.
$$

Because $G(\tau)$ is real, it is easy to see that

$$
\epsilon_r(-\omega) = [\epsilon_r(\omega^*)]^*.
$$

We also see that ϵ_r is an analytic function of ω in the upper half-plane. For $\Im\omega = 0$, convergence requires a discussion of $G(\tau)$ as $\tau \to \infty$. The factor $e^{i\omega\tau}$ will oscillate, which means that the worst situation is when $\omega = 0$, so that we consider

$$
\epsilon_r(0) = 1 + \int_0^\infty G(\tau)\, d\tau.
$$

In the case of a conductor, there is a pole at $\omega = 0$; ϵ_r behaves like $i\frac{\sigma}{\omega}$ as $\omega \to 0$. This can easily be accommodated if necessary, but setting this minor complication aside, the analyticity of $\epsilon_r(\omega)$ may be extended to $\Im\omega \geq 0$. The Cauchy theorem then allows us to write (for any z inside C)

$$
\epsilon_r(z) = 1 + \frac{1}{2\pi i}\oint_C \frac{[\epsilon_r(\omega') - 1]}{\omega' - z}\, d\omega',
$$

in which the contour of integration can be taken to be the familiar contour made up from the real axis and a semicircle at infinity in the upper half-plane. The arc at infinity makes no contribution to the integral, because $\epsilon_r - 1 \to 0$ sufficiently fast, and one has, for any z with $\Im z > 0$,

$$\epsilon_r(z) = 1 + \frac{1}{2\pi i} \int_{-\infty}^{\infty} \epsilon_r(\omega') - \frac{1}{\omega' - z} \, d\omega'.$$

We take $z = \omega + i\epsilon$, $\epsilon > 0$, and deduce

$$\epsilon_r(\omega) = \lim_{\epsilon \to 0} \left[1 + \frac{1}{2\pi i} \int_{-\infty}^{\infty} \frac{\epsilon_r(\omega') - 1}{\omega' - \omega - i\epsilon} \, d\omega' \right].$$

If we use

$$\lim_{\epsilon \to 0} \int_{-\infty}^{\infty} \frac{f(x)}{x - x_0 - i\epsilon} \, dx = P \int_{-\infty}^{\infty} \frac{f(x)}{x - x_0} + i\pi f(x_0)$$

(where $P\int$ denotes the Cauchy principal value integral), this gives

$$\epsilon_r(\omega) = 1 + \frac{1}{\pi i} P \int_{-\infty}^{\infty} \frac{\epsilon_r(\omega') - 1}{\omega' - \omega} \, d\omega',$$

or, separating real and imaginary parts,

$$\Re\epsilon_r(\omega) = 1 + \frac{1}{\pi} P \int_{-\infty}^{\infty} \frac{\Im\epsilon_r(\omega')}{\omega' - \omega} \, d\omega', \tag{12.9}$$

$$\Im\epsilon_r(\omega) = -\frac{1}{\pi} P \int_{-\infty}^{\infty} \frac{\Re\epsilon_r(\omega') - 1}{\omega' - \omega} \, d\omega'. \tag{12.10}$$

If we use also the previous result on ϵ_r^* which amounts to

$$\Re\epsilon_r(-\omega) = \Re\epsilon_r(\omega),$$
$$\Im\epsilon_r(-\omega) = -\Im\epsilon_r(\omega),$$

there follow the very important *Kramers-Kronig relations* [6]

$$\Re\epsilon_r(\omega) = 1 + \frac{2}{\pi} P \int_0^{\infty} \frac{\omega' \Im\epsilon_r(\omega')}{\omega'^2 - \omega^2} \, d\omega',$$

$$\Im\epsilon_r(\omega) = -\frac{2\omega}{\pi} P \int_0^{\infty} \frac{\Re\epsilon_r(\omega') - 1}{\omega'^2 - \omega^2} \, d\omega', \tag{12.11}$$

which had an important influence on the development of quantum mechanics, and which in their generalized form continue to be of great importance in studies of high-energy physics. Because in the optical case they give relations between the real part of ϵ_r, which determines *dispersion,* and the imaginary part, which determines *absorption*, these relations and their generalizations are called *dispersion relations*; they are a consequence of *analyticity* (here of ϵ_r), which in turn is intimately related to the *causality* of the physical process studied.

---

[6]Hendrik Anthony Kramers (1894–1952) and Ralph Kronig (1904–1995).

12.3 The Optical Theorem

In the chapter on scattering, we introduced the scattering amplitude and scattering cross-section, but in the context of that discussion we made some approximations appropriate for a small scatterer. What we want to derive now is a very general exact result that holds without approximation. This is the *optical theorem*, first encountered as its name suggests in connection with optics, but in fact also true for other scattering phenomena, and also indeed in quantum mechanics. It relates the imaginary part of the forward scattering amplitude to the total cross-section.

The total cross-section is the rate at which energy is removed from the incident wave divided by the flux of energy in the incident wave.[7] Without restriction to a small scatterer, we do still ask that the scatterer shall be localized, which allows us to maintain that, outside of the scattering region, the fields are *free*, and this will hold both for the incident wave and for the outgoing, scattered, wave. We will again consider a definite frequency $\omega = kc$ for the incident wave, a plane wave with wave-vector $\mathbf{k}_0$ and polarization direction $\boldsymbol{\epsilon}_0$. And we will use a Lorenz gauge which is also a Coulomb gauge, with scalar potential $\Phi = 0$, which means that the vector potential is *transverse*. The incident wave then has vector potential

$$\mathbf{A}_0(\mathbf{x}) = A_0 \boldsymbol{\epsilon}_0 e^{i\mathbf{k}_0 \cdot \mathbf{x}}.$$

We also have

$$\Box \mathbf{A}_0 = 0, \tag{12.12}$$
$$\mathbf{B}_0 = i\mathbf{k}_0 \times \mathbf{A}_0, \tag{12.13}$$
$$\mathbf{E}_0 = ikc\mathbf{A}_0. \tag{12.14}$$

Because of the scatterer, the vector potential is now $\mathbf{A} = \mathbf{A}_0 + \mathbf{A}_s$, where the additional term $\mathbf{A}_s$ represents the scattered wave. From Maxwell's equations it follows that the vector potential satisfies

$$\Box \mathbf{A} = \mu_0 (\mathbf{j} + \boldsymbol{\nabla} \times \mathbf{M} + \dot{\mathbf{P}}),$$

where we will set the current $\mathbf{j}$ to zero, since we are interested in scattering, not the radiation driven by a current provided from "outside"; the electric and magnetic polarizations ($\mathbf{P}$ and $\mathbf{M}$), respectively, are however those *induced* by the incident wave, and indeed it is they that generate the scattered wave, as is seen from the equation

$$\Box \mathbf{A}_s = \mu_0 (\boldsymbol{\nabla} \times \mathbf{M} + \dot{\mathbf{P}}), \tag{12.15}$$

and what follows from it on applying the Green's fuction approach, namely,

$$\mathbf{A}_s(\mathbf{x}) = \frac{\mu_0}{4\pi} \int d^3y \, G_k(\mathbf{x}, \mathbf{y})[\boldsymbol{\nabla} \times \mathbf{M} - i\omega\mathbf{P}]_{\mathbf{y}}). \tag{12.16}$$

---

[7]See footnote 2 in Chapter 11.

To define the scattering amplitude, we need to consider this contribution to the potential at large distances, so we may use

$$\mathbf{A}_s(\mathbf{x}) \sim \frac{e^{ikr}}{r} \frac{\mu_0}{4\pi} \int d^3y \, e^{-i\mathbf{k}\cdot\mathbf{y}} [\boldsymbol{\nabla} \times \mathbf{M} - ick\mathbf{P}]_\mathbf{y})$$

$$= A_0 \frac{e^{ikr}}{r} \mathbf{F}. \tag{12.17}$$

The integration may be restricted to any region large enough to enclose the scatterer since the integrand vanishes in the free space outside the scatterer. The *scattering amplitude F* is defined by

$$F(\mathbf{k}, \boldsymbol{\epsilon}; \mathbf{k}_0, \boldsymbol{\epsilon}_0) = \boldsymbol{\epsilon}^* \cdot \mathbf{F}. \tag{12.18}$$

The differential cross-section for scattered radiation with wave-vector $\mathbf{k}$ and polarisation $\boldsymbol{\epsilon}$ is then given by

$$\frac{d\sigma}{d\Omega} = |F(\mathbf{k}, \boldsymbol{\epsilon}; \mathbf{k}_0, \boldsymbol{\epsilon}_0)|^2. \tag{12.19}$$

The optical theorem relates the imaginary part of the forward scattering amplitude $F(\mathbf{k}, \boldsymbol{\epsilon}; \mathbf{k}, \boldsymbol{\epsilon})$ to the *total* cross-section. Recall that the total cross-section is the rate at which energy is removed from the incident wave divided by the flux of energy in the incident wave.

With $\mathbf{A}_0$ as given, since we have $\mathbf{E}_0 = ick\mathbf{A}_0 = ickA_0\boldsymbol{\epsilon}_0 \exp(i\mathbf{k}_0 \cdot \mathbf{x})$ and $\mathbf{B}_0 = i\mathbf{k}_0 \times \mathbf{A}_0 = iA_0\mathbf{k}_0 \times \boldsymbol{\epsilon}_0 \exp(i\mathbf{k}_0 \cdot \mathbf{x})$, the (time-averaged) incident flux is $ck^2|A_0|^2/2\mu_0$. The rate of flow of energy into the scatterer, given by integrating the Poynting vector across any surface S surrounding the scatterer, is

$$P_{\text{abs}} = -\frac{1}{2\mu_0} \int_S \mathbf{n} \cdot \Re[\mathbf{E} \times \mathbf{B}^*] \, dS,$$

where $\mathbf{n}$ is the outwards normal so that there is a minus sign for the *inwards* flow of energy, and the factor of one-half is for time-averaging. In addition to this absorbed power, there is power scattered out of the incident beam, given by

$$P_{\text{scattd}} = \frac{1}{2\mu_0} \int_S \mathbf{n} \cdot \Re[\mathbf{E}_s \times \mathbf{B}_s^*] \, dS.$$

With $\mathbf{E} = \mathbf{E}_0 + \mathbf{E}_s$, $\mathbf{B} = \mathbf{B}_0 + \mathbf{B}_s$ this gives

$$P = P_{\text{abs}} + P_{\text{scattd}}$$

$$= -\frac{1}{2\mu_0} \int_S \mathbf{n} \cdot \Re[\mathbf{E}_s \times \mathbf{B}_0^* + \mathbf{E}_0^* \times \mathbf{B}_s] \, dS$$

$$= \frac{1}{2\mu_0} \int_S \mathbf{n} \cdot \Re[iA_0^* e^{-i\mathbf{k}_0\cdot\mathbf{y}}(\mathbf{E}_s \times (\mathbf{k}_0 \times \boldsymbol{\epsilon}_0^*) + ck\boldsymbol{\epsilon}_0^* \times \mathbf{B}_s)] \, dS$$

$$= \frac{1}{2\mu_0} \Im \left\{ A_0^* \boldsymbol{\epsilon}_0^* \cdot \int_S e^{-i\mathbf{k}_0\cdot\mathbf{y}} [ck\mathbf{n} \times \mathbf{B}_s - (\mathbf{n} \times \mathbf{E}_s) \times \mathbf{k}_0] \, dS \right\}. \tag{12.20}$$

Hence we have for the total cross-section σ_{tot}, which is P divided by the incident flux,

$$\sigma_{\text{tot}} = \frac{1}{ck^2}\Im\left\{ A_0^{-1}\boldsymbol{\epsilon}_0^* \cdot \int_S e^{-i\mathbf{k}_0\cdot\mathbf{y}}[ck\,\mathbf{n}\times\mathbf{B}_s - (\mathbf{n}\times\mathbf{E}_s)\times\mathbf{k}_0]\,dS \right\}. \quad (12.21)$$

Now the scattering amplitude is

$$F = \frac{1}{A_0}\frac{\mu_0}{4\pi}\boldsymbol{\epsilon}^* \cdot \int d^3y\, e^{-i\mathbf{k}\cdot\mathbf{y}}(\boldsymbol{\nabla}\times\mathbf{M}+\dot{\mathbf{P}})_{\mathbf{y}},$$

and the integral in this expression, which is taken over any volume enclosing the scatterer, can be expressed by a succession of vector identities as follows:

$$\mu_0\int d^3y\, e^{-i\mathbf{k}\cdot\mathbf{y}}(\boldsymbol{\nabla}\times\mathbf{M}+\dot{\mathbf{P}})(\mathbf{y})$$

$$= \int d^3y\, e^{-i\mathbf{k}\cdot\mathbf{y}}\Box\mathbf{A}(\mathbf{y})$$

$$= \int d^3y\, e^{-i\mathbf{k}\cdot\mathbf{y}}\Box\mathbf{A}_s(\mathbf{y}) \qquad (\Box\mathbf{A}_{\text{in}}=0)$$

$$= \int d^3y\, e^{-i\mathbf{k}\cdot\mathbf{y}}(-k^2-\nabla^2)\mathbf{A}_s(\mathbf{y})$$

$$= \int d^3y\,[\mathbf{A}_s\nabla^2 e^{-i\mathbf{k}\cdot\mathbf{y}} - e^{-i\mathbf{k}\cdot\mathbf{y}}\nabla^2\mathbf{A}_s]$$

$$= \int_S [\mathbf{A}_s\,(\mathbf{n}\cdot\boldsymbol{\nabla})e^{-i\mathbf{k}\cdot\mathbf{y}} - e^{-i\mathbf{k}\cdot\mathbf{y}}\,(\mathbf{n}\cdot\boldsymbol{\nabla})\mathbf{A}_s]\,dS \qquad \text{(Green's theorem)}$$

$$= \int_S [2\mathbf{A}_s\,(\mathbf{n}\cdot\boldsymbol{\nabla})e^{-i\mathbf{k}\cdot\mathbf{y}} - (\mathbf{n}\cdot\boldsymbol{\nabla})(e^{-i\mathbf{k}\cdot\mathbf{y}}\mathbf{A}_s)]\,dS$$

$$= \int_S 2\mathbf{A}_s\,(\mathbf{n}\cdot\boldsymbol{\nabla})e^{-i\mathbf{k}\cdot\mathbf{y}}\,dS - \int d^3y\,\nabla^2(\mathbf{A}_s e^{-i\mathbf{k}\cdot\mathbf{y}}) \qquad \text{(divergence theorem)}$$

$$= \int_S 2\mathbf{A}_s\,(\mathbf{n}\cdot\boldsymbol{\nabla})e^{-i\mathbf{k}\cdot\mathbf{y}}\,dS - \int d^3y\,[\boldsymbol{\nabla}(\boldsymbol{\nabla}\cdot(\mathbf{A}_s e^{-i\mathbf{k}\cdot\mathbf{y}})) - \boldsymbol{\nabla}\times(\boldsymbol{\nabla}\times(\mathbf{A}_s e^{-i\mathbf{k}\cdot\mathbf{y}}))]$$

$$= \int_S [2\mathbf{A}_s\,(\mathbf{n}\cdot\boldsymbol{\nabla})e^{-i\mathbf{k}\cdot\mathbf{y}} - \mathbf{n}\,\boldsymbol{\nabla}\cdot(\mathbf{A}_s e^{-i\mathbf{k}\cdot\mathbf{y}}) + \mathbf{n}\times(\boldsymbol{\nabla}\times(\mathbf{A}_s e^{-i\mathbf{k}\cdot\mathbf{y}}))]\,dS \qquad \text{(Gree}$$

$$= \int_S [2\mathbf{A}_s\,(\mathbf{n}\cdot\boldsymbol{\nabla})e^{-i\mathbf{k}\cdot\mathbf{y}} - \mathbf{n}\,e^{-i\mathbf{k}\cdot\mathbf{y}}(\boldsymbol{\nabla}\cdot\mathbf{A}_s) - \mathbf{n}\,(\mathbf{A}_s\cdot\boldsymbol{\nabla})e^{-i\mathbf{k}\cdot\mathbf{y}} + e^{-i\mathbf{k}\cdot\mathbf{y}}\mathbf{n}\times(\boldsymbol{\nabla}\times$$
$$\qquad - \mathbf{A}_s\,(\mathbf{n}\cdot\boldsymbol{\nabla})e^{-i\mathbf{k}\cdot\mathbf{y}} + (\mathbf{n}\cdot\mathbf{A}_s)\boldsymbol{\nabla}e^{-i\mathbf{k}\cdot\mathbf{y}}]\,dS$$

$$= \int_S [\mathbf{A}_s\,(\mathbf{n}\cdot\boldsymbol{\nabla})e^{-i\mathbf{k}\cdot\mathbf{y}} + e^{-i\mathbf{k}\cdot\mathbf{y}}\mathbf{n}\times(\boldsymbol{\nabla}\times\mathbf{A}_s) + (\mathbf{n}\cdot\mathbf{A}_s)\boldsymbol{\nabla}e^{-i\mathbf{k}\cdot\mathbf{y}}$$
$$\qquad - \mathbf{n}\,e^{-i\mathbf{k}\cdot\mathbf{y}}(\boldsymbol{\nabla}\cdot\mathbf{A}_s) - \mathbf{n}\,(\mathbf{A}_s\cdot\boldsymbol{\nabla})e^{-i\mathbf{k}\cdot\mathbf{y}}]\,dS$$

$$= \int_S [-i\mathbf{A}_s\,(\mathbf{n}\cdot\mathbf{k}) + \mathbf{n}\times(\boldsymbol{\nabla}\times\mathbf{A}_s) - i\mathbf{k}\,(\mathbf{n}\cdot\mathbf{A}_s) - \mathbf{n}\,(\boldsymbol{\nabla}\cdot\mathbf{A}_s) + i\mathbf{n}\,(\mathbf{k}\cdot\mathbf{A}_s)]e^{-i\mathbf{k}\cdot\mathbf{y}}$$

We may now impose the radiation gauge condition $\boldsymbol{\nabla}\cdot\mathbf{A}=0$, and use

$$ikc\mathbf{A}_s = \mathbf{E}_s \quad\text{and}\quad \boldsymbol{\nabla}\times\mathbf{A}_s = \mathbf{B}_s,$$

as well as the fact that the scattered radiation is in the direction of $\mathbf{n}$ and is transverse so that $\mathbf{n} \cdot \mathbf{E}_s = 0$, to write this as

$$\frac{1}{kc} \int_S [-\mathbf{E}_s (\mathbf{n} \cdot \mathbf{k}) + kc\mathbf{n} \times \mathbf{B}_s + \mathbf{n} (\mathbf{k} \cdot \mathbf{E}_s)]e^{-i\mathbf{k} \cdot \mathbf{y}}\, dS.$$

This then gives for the scattering amplitude

$$\begin{aligned}
F(\mathbf{k}, \boldsymbol{\epsilon}; \mathbf{k}_0, \boldsymbol{\epsilon}_0) &= \frac{1}{4\pi kcA_0} \boldsymbol{\epsilon}^* \cdot \int_S [-\mathbf{E}_s (\mathbf{n} \cdot \mathbf{k}) + kc\mathbf{n} \times \mathbf{B}_s + \mathbf{n} (\mathbf{k} \cdot \mathbf{E}_s)]e^{-i\mathbf{k} \cdot \mathbf{y}}\, dS \\
&= \frac{1}{4\pi kcA_0} \boldsymbol{\epsilon}^* \cdot \int_S [kc\mathbf{n} \times \mathbf{B}_s - (\mathbf{n} \times \mathbf{E}_s) \times \mathbf{k}]e^{-i\mathbf{k} \cdot \mathbf{y}}\, dS. \quad (12.22)
\end{aligned}$$

Comparing this with the previously obtained expression for the total cross-section, we arrive at

$$\sigma_{\text{tot}} = \frac{4\pi}{k} \Im F(\mathbf{k}_0, \boldsymbol{\epsilon}_0; \mathbf{k}_0, \boldsymbol{\epsilon}_0), \qquad (12.23)$$

which is the optical theorem, expressing the total cross-section as $4\pi/k$ times the imaginary part of the forward-scattering amplitude.

Epilogue

In 1862 Maxwell was Professor of Natural Philosophy in King's College, London. His research on electricity and magnetism, especially in developing a theoretical basis for an explanation of the experimental work of Faraday, had led him to contrive a model in which he had "shown how the forces acting between magnets, electric currents, and matter capable of magnetic induction may be accounted for on the hypothesis of the magnetic field being occupied with innumerable vortices of revolving matter, their axes coinciding with the direction of the magnetic force at every point of the field."[1] He was struck by the observation:

> The velocity of transverse undulations in our hypothetical medium, calculated from the electro-magnetic experiments of MM. Kohlrausch and Weber, agrees so exactly with the velocity of light calculated from the optical experiments of M. Fizeau, that we can scarcely avoid the inference that light consists in the transverse undulations of the same medium which is the cause of electric and magnetic phenomena.

This ingenious mechanical model of cellular vortices was to be superceded two years later by his "Dynamical Theory of the Electromagnetic Field."[2] But his conclusion still survived in his new theory of electromagnetic waves; he was able to calculate the speed of propagation of the waves (as we have done in Chapter 3) and so to state again:

> This velocity is so nearly that of light, that it seems we have strong reason to conclude that light itself (including radiant heat, and other raiations if any) is an electromagnetic disturbance in the form of waves propagated through the electromagnetic field according to electromagnetic laws.

There was no strong reason to hesitate in declaring that light was a manifestation of electromagnetism, because it was accepted that light propagated as waves. However, the alternative hypothesis that light was corpuscular, was one to which Newton inclined. But it was for him a hypothesis, as is clear from the speculative context in which it was formulated as one of his "Questions" that conclude the *Opticks*:

---

[1]In Part III of J. C. Maxwell, " On Physical Lines of Force," *Phil. Mag.* April and May, 12–24 (1861).

[2]*Phil. Trans. R. Soc.* **155**, 459–512 (1865). Read 8 Dec. 1864.

Qu. 29. Are not the Rays of Light very small Bodies emitted from shining Substances?

He gives arguments in support of this hypothesis:

> For such Bodies will pass through uniform Mediums in right Lines without bending into the Shadow, which is the Nature of Rays of Light. They will also be capable of several Properties, and be able to conserve their Properties unchanged in passing through several Mediums, which is another Condition of Rays of Light. [...]

He was vehement in his critique of a wave theory of light.

> *Qu.* 28. Are not all Hypotheses erroneous, in which Light is supposed to consist in Pressions or Motion propagated through a fluid Medium? [...] If it consisted in Pression propagated to all distances in an instant, it would require an infinite force every moment, in every shining Particle to generate that Motion. And if it consisted in Pression or Motion, propagated either in an instant or in time, it would bend into the Shadow. For Pression or Motion cannot be propagated in a Fluid in right Lines, beyond an Obstacle which stops part of the Motion, but will bend and spread every way into the quiscent Medium which lies beyond the Obstacle.

So for Newton, the lack of what we now call diffraction is an argument against the wave theory. But Newton was well aware of the work of Grimaldi[3] which described diffraction fringes outside the geometric shadow of a sharp edge.

Newton was also familiar with the birefringence of "Island Crystal" (Iceland Spar), which we now understand as a consequence of different refractive indices for left- and right-circular polarizations of light. Since for him waves were *longitudinal*, like sound or water waves, he was unable to consider the possibility of polarization. So birefringence was incompatible with a wave theory.[4] Instead he suggests that the corpuscles of light have "sides."

> Every Ray of Light has therefore two opposite Sides, originally endued with a Property on which the unusual Refraction depends, and the other two opposite Sides not endued with that Property.

For Newton diffraction and refraction result from a force acting on the corpuscles (particles) of light.

---

[3]F. M. Grimaldi *Physico-Mathesis de Lumine Coloribus et Iride* (Bologna, 1665). Indeed, the third book of the *Opticks* begins with an extensive discussion of this work, and details careful experiments that he performed to extend Grimaldi's obsevations, leading him to ask in Query 3: Are not the Rays of Light in passing by the sides of Bodies, bent several times backwards and forwards, with a motion like that of an Eel? and do not the three Fringes of colour'd Light above-mention'd arise from three such bendings?

[4]He gives an excellent account of the phenomenon, but rejected the wave-theoretic explanation of Christiaan Huygens (1629–1695) (in *Traité de la Lumière* (Pieter van der Aa, Leiden, 1690) which, as noted by Huygens himself, fails fully to account for what is observed.

Query 1. Do not Bodies act upon Light at a distance, and by their action bend its Rays; and is not this action (*cæteris paribus*) strongest at the least distance?

Qu. 2. Do not the Rays which differ in Refrangibility differ also in Flexibility; and are they not by their different Inflexions separated from one another, so after separation to make the Colours in the three Fringes above described. [...]

Qu. 19. Doth not the Refraction of Light proceed from the different density of this Æthereal Medium in different places, the Light receding away from the denser part of the Medium? And is not the density thereof greater in free and open Spaces void of Air and other grosser Bodies, than within the Pores of Water, Glass, Crystal, Gems, and other compact Bodies? For when Light passes through Glass or Crystal, and falling very obliquely upon the further Surface thereof is totally reflected, the total Reflexion ought to proceed rather from the density and vigour of the Medium without and beyond the Glass, than from the rarity and weakness thereof.

This "Æthereal Medium" is related to the "lumeniferous æther" of Huygens, needed to carry the waves in his theory of the propagation of light, an idea which persisted into the beginning of the twentieth century. For Newton it exerts a *repulsive* force on the particles of light, and since he supposes it to be *denser* in a vacuum than in glass, for example, the net result is that light goes *faster* in glass than in a vacuum. and it is the net repulsion on the particles of light incident from glass on an interface with air that can lead to total internal reflection. But although Newton accepted the idea of an Æther, it was for him not the carrier of light waves, although it did exert a force on his light corpuscles. And as we shall see, it played a role in the strange and tortuous explanation for one of the phenomena that bears his name—that of *Newton's rings.*

A large part of Book Two of the *Opticks* is devoted to a description of his observations of the alternating rings of dark and light that are seen when a the convex surface of a plano-convex lens is pressed against a planar sheet of glass, and to the analogous phenomena when light is viewed through a narrow wedge of air between two sheets of glass in contact at one edge but separated by a small gap at the opposite edge. He also describes the colors seen in bubbles. The present-day explanation in terms of interference of waves was of course not open to Newton with his corpuscular theory. Instead he had to invent an idea that is perilously close to one of waves. So in Proposition XII of Book Two:

Every Ray of Light in its passage through any refracting Surface is put into a certain transient Constitution or State, which in the progress of the Ray returns at equal Intervals, and disposes the Ray at every return to be easily transmitted throught the next refracting Surface, and between the returns to be easily reflected by it.

And in the following Definition:

The returns of the disposition of any Ray to be reflected I will call its *Fits of easy Reflexion*, and those of its disposition to be transmitted its *Fits of easy Transmission*, and the space it passes between every return and the next return, the *Interval of its Fits*.

It is vibrations in the æther that are responsible for these Fits.

> *Qu. 29.* [...] Nothing more is requisite for putting the Rays of Light into Fits of easy Reflexion and easy Transmission, than that they be small Bodies which by their attractive Powers, or some other Force, stir up Vibrations in what thy act upon, which Vibrations being swifter than the Rays, overtake them successively, and agitate them so as by turns to increase and decrease their Velocities, and thereby put them into those Fits.

What I find truly remarkable is that Newton determines the "Interval of the Fits" from his measurements of the diameters of the rings, giving the result for light in the yellow-orange part of the spectrum "$\frac{1}{89000}$th part of an Inch" or approximately 290 nm. And this is an excellent determination of what is half of the wavelength in the wave theory, such as might be made in an undergraduate class using the same measurements and essentially the same calculations as used by Newton. So no waves, but a good determination of the wavelength.

Was Newton being stubborn, obstinate, in his rejection of the wave theory, or was it based on the kind of philosophic prejudices that he claimed to abjure?

It needed new physical insights, new theoretical understanding, and most of all, new experimental results, to tilt the balance emphatically in favor of the wave theory. Building on the wave approach of Huygens, Thomas Young (1773–1829) argued by analogy with the *transverse* waves on the surface of water that he studied with a "ripple tank" that light waves might be transverse rather than longitudinal, as had been supposed by Newton and indeed by Huygens. Since there are two independent transverse directions for a wave propagating in space, the objection to the wave theory voiced by Newton in his account of birefringence was removed, and the two independent polarizations of light rays become possible. But of greater significance was Young's recognition that what he saw on the surface of his ripple tank, namely, the interference of waves produced by two coherent sources, might have an analog in light. And so he devised and performed his celebrated *two-slit experiment*, which brilliantly demonstrates interference effects. He also was able to use the construction deployed by Huygens for describing successive wave-fronts to explain the results of his experiments, and also other phenomena that would now be regarded as direct evidence in favor of the wave theory, including diffraction—and Newton's rings.

There was still to come the *experimentum crucis*—the crucial experiment to decide between particles and waves. This hinges on the speed of light in a dispersive medium, which in the particle theory (as posited by Newton) is greater than that in a vacuum by a factor that can be identified with the refractive index, while in the wave theory (as required by Huygens) it is less than that in a vacuum by the same factor. The speed of light in a vacuum was known to Newton from the brilliant observations by Ole Römer of the orbits of Io, one of Jupiter's moons. He quotes the time for light to travel from the Sun to Earth as eight minutes. But how to measure the speed of light in a medium?

Augustin Jean Fresnel (1788–1827) extended and refined Young's work, especially on diffraction and polarization. He argued that the speed of light could be different in a moving medium from one at rest because the aether might be dragged along in the moving medium. This aether drag adjustment allowed him to explain the null results obtained by Françoise Arago on the seasonal and temporal changes he had expected in stellar aberration and gave further support to the wave theory. Hyppolyte Fizeau (1819-1896) devised an experiment to measure the speed of light in flowing water that confirmed the prediction of Fresnel. This ingenious experiment passed light from a single source split by a half-silvered mirror so as to pass through two parallel tubes containing fast-flowing water, in such a way that in one tube the light was in the same direction as the water flow and in the other tube in the opposite direction. Bringing the two components back together generated interference fringes that were shifted as the speed of the water flow was increased. The relation between the shift of the fringes and the speed of the water gave the desired confirmation of Fresnel's aether drift. But does this not also confirm the existence of the aether entrained in the moving water?

The answer is negative, because Hendrik Lorentz (1853–1928) was able to show that the same result could be obtained without the aether hypothesis, a result which was regarded as being of great significance by Albert Einstein in his developing ideas on the theory of relativity.

And so we come to the twentieth century, a century rich in revolutionary ideas, of which the first was that of Max Planck (1858–1947). The problem presented by the deviation between experiment and the attempts by classical theory to explain the spectrum of *black body radiation* was in the air in the closing years of the nineteenth century. Planck's work was presented at a meeting of the Deutsche Physikalische Gesellschaft, the German Physical Society, in October 1900.[5] He postulated

> ...it is necessary to interpret [the energy of an oscillator] not as a continuous, infinitely divisible quantity, but as a discrete quantity composed of an integral number of finite equal parts. Let us call each such part the energy element ϵ.

And he then goes on to assert

> that the energy element ϵ must be proportional to the frequency ν, thus:
> $$\epsilon = h\nu.$$

With this assumption, he was able to derive his now celebrated formula for the black body spectrum, which to this day has survived unchallenged as providing complete agreement with experiment.[6] And thus was born quantum physics.

The significance of this rupture with classical physics was not immediately appreciated. Not until some five years had passed was another work of genius to

---

[5]Published in "On the Law of Distribution of Energy in the Normal Spectrum," *Ann. Phys.*, **4**, 553–563 (1901).

[6]Perhaps the most remarkable illustration of this is provided by the excellence of the fit between the observed spectrum of the cosmic background radiation and the Planckian formula.

provide confirmation of the quantum hypothesis. And this originated from an entirely different topic in the interaction of radiation with matter—the *photo-electric effect*. This had been discovered in 1902 by Philipp Lenard (1862–1947), who showed that, when metals were irradiated with ultraviolet radiation, they emitted electrons with an energy independent of the intensity of the ultraviolet radiation, but greater the shorter the wavelength. In one of the great papers[7] in his *annus mirabilis*, 1905, Albert Einstein gave an explanation of the photoelectric effect which treated the ultraviolet radiation as *corpuscular*, predicting a linear relation between the energy of the emitted electrons and the frequency of the ultraviolet radiation, with the Planck constant h as the slope of this linear relation. It took some two decades before these corpuscles, or quanta of light, were to be called *photons*.[8] Einstein's theory was at first rejected by Planck, among others, but was to be vindicated by experiments, including those in 1915 by Robert A. Millikan (1868-1953), and those in 1923 by Arthur Holly Compton (1892-1962), who went further by ascribing particle-like momentum as well as energy to the photons that scatter from electrons in the effect named for him.[9]

What emerges from these developments is that electromagnetic radiation, including of course light, exhibits both wavelike and particle-like properties. This *wave-particle duality* is basic to quantum physics. It achieves a reconciliation after more than two centuries between the opposing views of Huygens and Newton that neither could have thought possible.

There were several major advances in the development of quantum physics in the first quarter of the twentieth century, but it was still a mysterious blend of ideas borrowed from classical physics with some restrictions and "rules" imposed on them. An important example was of course the theory of atomic spectra elaborated by Niels Bohr (1885–1962) and his collaborators in Copenhagen. Among those who were attracted to Copenhagen to work with Bohr was Hendrik Kramers (1894–1952).[10] Another focus of quantum physics research was Göttingen, where Arnold Sommerfeld (1868–1951) and Max Born (1882–1970) were prominent professors. It was there that the young Werner Heisenberg (1901–1976), one of Sommerfeld's students, met Bohr. Encouraged by Bohr, Kramers and Heisenberg collaborated on an important paper[11] giving a dispersion formula for the cross-section for scattering of a photon by an electron in an atom.

Later that same year (1925), Heisenberg left Göttingen for the North Sea island of Helgoland, an island with very few trees, to escape from the pollen that had given him hay fever. He took with him the problem on which he was working, the calculation of the intensities of the lines in the spectrum of atomic hydrogen, trying to use the methods he had been developing with Kramers.

---

[7] "On a Heuristic Viewpoint Concerning the Production and Transformation of Light," *Ann. Phys.* **17**, 132-148 (1905).

[8] By Gilbert N. Lewis, *Nature* **118**, 874 (1926).

[9] Lenard (1905), Planck (1918), Einstein (1921), Millikan (1923), and Compton (1927) were all Nobel Prize winners.

[10] The dispersion relations derived by him and Ralph Kronig (1904–1995) date from 1926.

[11] H. A. Kramers and W. Heisenberg, *Z. Phys.* **48**, 15 (1925).

But this problem was too technically complicated, so he turned instead to the simpler one of an anharmonic oscillator. In his autobiographical book[12] he wrote

> It was about three o'clock at night when the final result of the calculation lay before me. At first I was deeply shaken. I was so excited that I could not think of sleep. So I left the house and awaited the sunrise on the top of a rock.

What he had found was that by using *noncommuting observables* he could solve the problem. This (perhaps in an overromanticized account) was how quantum physics gave birth to *quantum mechanics*.

The noncommuting variables introduced by Heisenberg in the breakthrough paper that resulted from his "holiday" in Helgoland[13] were recognized by his mentor Max Born and Born's research student Pascual Jordan (1902–1980) as objects known to mathematicians as matrices.[14] So Heisenberg's new quantum mechanics is often called matrix mechanics. The publishers sent the proofs of Heisenberg's paper to Ralph Fowler (1889–1944) in Cambridge, who in turn passed them to his research student Paul Dirac (1902-1984). Thus stimulated, Dirac wrote a remarkable paper introducing a very general mathematical formulation of the new theory.[15] Other developments followed very rapidly, including a "three-man paper"[16] which also developed the new theory in a more systematic way. The statistical basis of the interpretation of quantum mechanics was introduced by Born that same year.[17]

In Göttingen and Copenhagen, the focal points of research using the new matrix mechanics, a galaxy of talented, mainly young, researchers tackled problems that could not be addressed by classical physics. But at the same time another star was about to shine. Erwin Schrödinger (1887–1961) wrote from the University of Zürich a paper[18] that challenged the supremacy of matrix mechanics. This work (in fact completed in 1925, the year of the qauntum revolution) used ideas and mathematical techniques that were less "exotic" than matrices. As Schrödinger expressed it in the intoduction of his paper:

> In this paper I wish to consider, first, the simple case of the hydrogen atom (non-relativistic and unperturbed), and show that the customary quantum conditions can be replaced by another postulate, in which the notion of "whole numbers", merely as such, is not

---

[12] *Der Teil Und Das Ganze: Gespräche Im Umkreis Der Atomphysik* Translated from the German by Arnold J. Pomerans as *Physics and Beyond; Encounters and Conversations* (Harper and Row, London, 1971).

[13] "Über quantentheoretische Umdeutung kinematischer und mechanischer Beziehungen" ("Quantum Theoretical Reinterpretation of Kinematic and Mechanical Relations"), *Z. Phys.* **33**, 879–893 (1925), received 29 July 1925.

[14] M. Born and P. Jordan, "Zur Quantenmechanik," *Z. Phys.* **34**, 858–888 (1925).

[15] P. Dirac, "The Fundamental Equations of Quantum Mechanics," *Proc. R. Soc. Lond.* **A109**, 642–653 (1925).

[16] M. Born, W. Heisenberg, and P. Jordan, "Zur Quantenmechanik II," *Z. Phys.* **35**, 557–615 (1926).

[17] M. Born, "Zur Quantenmechanik der Stossvorgänge," *Z. Phys.* **37**, 863–867 (1926) and *ibid* **38**, 803–827 (1926).

[18] "Quantisierung als Eigenwertproblem," *Ann. Phys.* **79**, 361–376 (1926).

introduced. Rather when integralness does appear, it arises in the same natural way as it does in the case of the node-numbers of a vibrating string. The new conception is capable of generalization, and strikes, I believe, very deeply at the true nature of the quantum rules.

His *wave mechanics*, as he called it, was able to obtain the quantum numbers that had been used by Bohr to identify the energy levels of atomic systems, with the same success as matrix mechanics.

The tension between these two contrasting approaches was not resolved by the publication of another paper by Schrödinger in 1926,[19] although in it he proved that matrix mechanics and wave mechanics were mathematically equivalent.[20]

The antagonism between the two camps was centered on their very different physical interpretations of the mathematics. Schrödinger insisted on the "visualizability" of his waves: "I knew of [Heisenberg's] theory, of course, but I felt discouraged, not to say repelled, by the methods of transcendental algebra, which appeared difficult to me, and by the lack of visualizability." But Heisenberg was adamant in maintaining the view expressed already in his first paper on matrix mechanics: "The present paper seeks to establish a basis for theoretical quantum mechanics founded exclusively upon relationships between quantities which in principle are observable." Schrödinger's waves are not observable. The debate between the two camps was intense and often acrimonious. Echoes of this can still be heard, but with the passage of time it is generally accepted that the Schrödinger equation is more accessible and easier to use for many problems, while the matrix manipulation is best used in many others. With the

---

[19]E. Schrödinger "Über das Verhältnis der Heisenberg-Born-Jordanschen Quantenmechanik zu der meinen" ("On the Relationship of the Heisenberg-Born-Jordan Quantum Mechanics to Mine"), *Ann. Phys.* **79**, 734–756 (1926). Received 18 March 1926.

[20]The opening paragraphs of his paper are as follows: Considering the extraordinary differences between the starting-points and the concepts of Heisenberg's quantum mechanics and of the theory which has been designated as "undulatory" or "physical" mechanics, and has lately been described here, it is very strange that these two theories agree *with one another* with regard to the known facts, where they differ from the old quantum theory. I refer, in particular to the peculiar "half-integralness" which arises in connection with the oscillator and the rotator. That is really very remarkable, because starting-points, presentations, methods, and in fact the whole mathematical apparatus, seem fundamentally different. Above all, however, the departure from classical mechanics in the two theories seems to occur in diametrically opposed directions. In Heisenberg's work the classical variables are replaced by systems of discrete numerical quantities (matrices), which depend on a pair of integral indices, and are defined by *algebraic* equations. The authors themselves describe the theory as a "true theory of a discontinuum." On the other hand, wave mechanics shows just the reverse tendency; it is a step from classical point-mechanics towards a *continuum theory*. In place of a process described in terms of a finite number of dependent variables occurring in a finite number of total differential equations, we have a continuous *field-like* process in configuration space, which is governed by a single *partial* differential equation, derived from a principle of action. This principle, and this differential equation replace the equations of motion *and* the quantum conditions of the older "classical quantum theory."

In what follows the very intimate *inner connection* between Heisenbergs quantum mechanics and my wave mechanics will be disclosed. From the formal mathematical standpoint, one might well speak of the *identity* of the two theories.

confidence that they are equivalent, we may choose the best tools for the job at hand, and leave the rest to philosophers.

At first quantum mechanics was applied to problems involving nonrelativistic motion of particles (although Schrödinger's first attempt—already in 1925— at an equation for the electron in a hydrogen atom had been relativistic; he abandoned it in favor of the nonrelativistic equation because it gave wrong answers for the fine structure of the hydrogen spectrum). A few years later Dirac produced his celebrated relativistic equation for the electron[21] which predicted its spin and magnetic dipole moment, and—a little later—the positron, the electron's antiparticle. Dirac had already developed the quantum theory of the electromagnetic field,[22] and it was in this paper that the phrase *quantum electrodynamics* was first used. With the incorporation of Dirac's relativistic treatment of the electron this gave the framework for what is to this day the starting point for quantum electrodynamics—QED. The theory follows from a Lagrangian that is in essence the same as that used in the classical theory presented in Section 5.4. The passage from the classical theory to its quantum-mechanical counterpart may be made via the classical Hamiltonian and then by regarding the fields as operators in the same way that Heisenberg regarded the dynamical variables for particles as operators. The electromagnetic field is regarded as an operator that can effect the creation or annihilation of *quanta*, the particle-like manifestation of electromagnetic radiation—*photons*. The current with which the electromagnetic field interacts is constructed from quantum fields that are also operators that can create or annihilate particles—electrons and so forth.

The electromagnetic fields still propagate as waves, but, as operators, create or annihilate photons that have particle-like properties. So the quantum field theory resolves the old dilemma between these two contrasting designations. Light is neither just waves nor just corpuscular.

Now it is all very well to set out a recipe for a relativistic field theory with interactions as above. But making use of this recipe to calculate quantities accessible to experiment is not so easy! The interactions in QED are gratifyingly weak enough that a perturbation approach is realistic (their strength is governed by the dimensionless number known as the fine-structure constant, constructed from the charge on the electron, the speed of light, and the Planck constant; its value is close to $1/137$). But although the lowest-order calculations proceed in a straightforward fashion, and agree well with experiment, serious difficulties emerge when one tries to go to the next order of approximation in the perturbation expansion. The quantities to be calculated involve divergent integrals and frustrate application of the theory as it stood after some two decades of effort.

The modern theory of QED was born in the late 1940s as the confluence of independent research by two young Americans, Richard Feynman (1918–1988), Julian Schwinger (1918–1994), and a Japanese physicist Sin-Itiro Tomonaga

---

[21]P. A .M. Dirac, "The Quantum Theory of the Electron," *Proc. Roy. Soc. Lond.* **A117**, 610–624 (1928).

[22]P. A. M. Dirac, "The Quantum Theory of the Emission and Absorption of Radiation," *Proc. R. Soc. Lond.* **A114**, 243–265 (1927).

$(1906-1979)^{23}$ for which they were to share the 1965 Nobel Prize in Physics. As Freeman Dyson (1923–) wrote[24] while some of this the research was as yet unpublished: "The advantages of the Feynman theory are simplicity and ease of application, while those of Tomonaga-Schwinger are generality and theoretical completeness."[25] And in a paper[26] published a little later:

> The covariant quantum electrodynamics of Tomonaga,[27] Schwinger,[28] and Feynman[29] is used as the basis for a general treatment of scattering problems involving electrons, positrons, and photons. Scattering processes, including the creation and annihilation of particles, are completely described by the S matrix of Heisenberg. It is shown that the elements of this matrix can be calculated, by a consistent use of perturbation theory, to any desired order in the fine-structure constant.

QED is a covariant and gauge-invariant prescription for the calculation of observable physical quantities that can be compared with experiment. And the agreement is spectacular. For example the magnetic moment of the electron as calculated agrees with what is measured to one part in 10^{12}!

The success in taming the infinities that had plagued previous attempts to derive meaningful predictions beyond the lowest-order perturbation theory results depends critically on the gauge invariance that we encountered in Section 4.4. QED is the paradigm on which the so-called *standard model*[30] of high energy physics is built, a theory which incorporates the *strong interactions*,[31] and also the *weak interactions* through their unification with electromagnetism in the *electroweak* theory of Sheldon Glashow (1932–), Abdus Salam (1926–1996), and Steven Weinberg (1933–). These ideas of covariance and gauge invariance

---

[23] As he was to say in his Nobel lecture: This period, around 1946–1948, was soon after the second world war, and it was quite difficult in Japan to obtain information from abroad.

[24] F. J. Dyson, "The radiation theories of Tomonaga, Schwinger and Feynman," *Phys. Rev.* **75**, 486–502 (1949).

[25] In some ways this might be seen as analogous to the wave mechanics/matrix mechanics approaches to quantum mechanics—but without the acrimonious disputes between their originators.

[26] F. J. Dyson, "The S Matrix in Quantum Electrodynamics," *Phys. Rev.* **75**, 1736–1755 (1949).

[27] In a series of papers from "On a Relativistically Invariant Formulation of the Quantum Theory of Wave Fields," *Prog. Theor. Phys.* **1**, 27–42 (1946) to one with J. Robert Oppenheimer (1904–1967) "On Infinite Field Reactions in Quantum Field Theory," *Phys. Rev.* **74**, 224–225 (1948).

[28] From J. Schwinger, "On Quantum Electrodynamics and the Magnetic Moment of the Electron," *Phys. Rev.* **73**, 416–417 (1948) to "Quantum Electrodynamics III: The Electromagnetic Properties of the Electron—Radiative Corrections to Scattering," *Phys.Rev.* **76** 790–817 (1949).

[29] From R. P. Feynman "The Theory of Positrons," *Phys. Rev.* **76**, 749–759 (1949), to "Mathematical Formulation of the Quantum Theory of Electromagnetic Interaction," *Phys. Rev.* **80**, 440–457 (1950).

[30] See footnote 6 of Chapter 5.

[31] As QCD or quantum chromodynamics: *quarks*, the fundamental entities in the strong interactions carry "color," analogous to charge, and interact by the exchange of *gluons*, analogous to photons.

feature prominently in the continuing research on superstring theory and M-theory.

Feynman has called QED "the jewel of physics." Its remarkable and enduring success is, however, based on perturbation theory, and a word of caution is appropriate. Dyson[32] gives a tentative argument that the perturbation expansion is divergent, albeit asymptotically. But he concludes on a more positive note with an encouragement for future research:

> Experimentally we know that the world contains one group of phenomena which is accurately in agreement with the results of quantum electrodynamics [and one might add, the other sectors of the standard model], and another group of phenomena which is not understood at all. We need to develop new physical ideas to understand the second group, and still we cannot abandon the theory which successfully accounts for the first.

---

[32]F. J. Dyson, "Divergence of Perturbation Theory in Quantum Electrodynamics," *Phys. Rev.* **85**, 631–632 (1952).

Index

acceleration field, 91
action, 49
    principle of stationary action, 50
action-at-a-distance, 5
    problematic for Newton, 5
    also for Faraday and Maxwell, 6
amp
    unit of electric current, 2
    definition of, 32
Ampère, André-Marie, 2, 28
attenuation coefficient, 133

Biot-Savart law, 29, 72, 87
Bohr, Niels, 154–156
Born, Max
    Born approximation, 126–128
boundary conditions
    at interface, 111–114
    for Green's function, 79
    outgoing, 83, 88
Bremsstrahlung, 105
Brewster angle, 116

Čerenkov radiation, 116–119
charge density, 7, 27, 37
charge-current density, 37–38, 87
charged oscillator, 94
    Poynting vector, 94
Clausius-Mossotti relation, 131, 134
colliders (LHC, LEP), 104
conductivity, 8, 109, 140–141
conservation
    connection with gauge invariance,
        56–60
    of angular momentum, 64
    of charge, 29
    of energy, 50, 62–64
constitutive relations, 8, 25–26, 109

continuity equation, 38
contravariant vector, 16
coulomb
    definition of, 31–32
Coulomb, Charles Augustin, 2
    Coulomb field, 71–72
    Coulomb gauge, 42, 145
    Coulomb potential, 100
    Coulomb's Law, 27
    Coulomb's law, 2, 31, 87
    coulomb, unit of electric charge, 2
covariant derivative, 59
covariant vector, 16
critical opalescence, 133–135
current density, 37
curvature drift, 76

Davy, Humphrey, 2
delta function, 80
Diamond light source, 104
dielectric, 29–33
dipole moment, 25–26, 100–103, 127,
        128
dipole radiation, 94, 100–102, 122
Dirac, Paul Adrien Maurice, 28, 155,
        157
dispersion, 110, 137
    anomalous, 139
    dispersion relations, 142
displacement current, 7, 29–30
Drude model, 140–142
Dyson, Freeman, 158–159

Einsein, Albert
    The Principle of Relativity, 35
Einstein, Albert, 4, 13, 137, 153–154
    Einstein tensor, 19
    Einstein's equations, 19

summation convention, 16
Einstein-Smoluchowski equation, 135
electric charge, 1
electric constant ϵ_0, 26
electric current, 2
electric current density, 7
electromagnetic field, 4, 37
    Lorentz transformation law, 44
electromagnetic waves, 4, *see* Hertz, Heinrich Rudolf, 33–34
energy
    in a charged capacitor, 30, 32–33
    in a solenoid carrying a current, 33
    kinetic, 50–51, 54
    potential, 54–56
    relativistic energy of a particle, 22
energy density, electric, 33
energy density, magnetic, 33

farad, unit of capacitance, 2
Faraday, Michael, 2
    Law of Induction, 30
Feynman, Richard, 157–159
field
    displacement field **D**, 25
    electric field **E**, 25
    magnetic field **H**, 25
    magnetic induction **B**, 25
field theory, 3
field-strength tensor, 43
FitzGerald, George, 9, 19
Fizeau, Hyppolite, 153
four-vector
    covariant, contravariant, 16
Franklin, Benjamin, 1
Fresnel, Augustin Jean, 153

gauge invariance, 55–58, 158
gauge transformation, 40
Gauss, Carl Friedrich, 2
    Gauss's divergence theorem, 12
gauss, unit of magnetic induction, 2
Gilbert, William
    gilbert, unit for magnetic potential, 1
gradient drift, 75

Green, George
    covariant Green's function, 84–85
    Green's function, 81, 125, 145
    Green's theorem, 147
    retarded Green's function, 84
group velocity, 140

Hamilton, William Rowan
    Hamilton's equations, 53
    Hamiltonian density, 62
    Hamiltonian for charged particle, 53–54
    Hamiltonian for electromagnetic field, 57
harmonic oscillator, 68–69
Heaviside, Oliver
    Maxwell's equations using vector notation, 8
Heisenberg, Werner, 154–158
Helmholtz function, 80
Henry, Joseph, 2
henry, unit of inductance, 2
Hertz, Heinreich Rudolph
    confirms Maxwell's prediction that electromagnetic waves propagate with same velocity as light, 6
Hertz, Heinrich Rudolph, 6
Huygens, Christiaan, 150–154

induction
    magnetic, 2, 25–26
induction, Faraday's experiments on, 2
ionosphere, 124, 140

Kramers, Hendrik Anthony, 144, 154
Kramers-Kronig relations, 144
Kronig, Ralph, 144

Lagrange, Joseph-Louis, 49
    Lagrange's equations, 49–50
    Lagrangian, 49
    Lagrangian density, 55, 61
    Lagrangian for charged particle, 51–53
    Lagrangian for electromagnetic field, 54–56

Larmor, Joseph
Larmor formula, 96
Lenard, Philipp Eduard Anton, 154
Liénard, Alfried Marie
Liénard formula, 96
Liénard-Wiechert potentials, 89–91
linear accelerators
radiation from, 105
lines of force, 3–6, 61, 66
Lorentz boost, 14
Lorentz, Hendrik, 19, 153
Lorentz group, 17
Lorentz-FitzGerald contraction, 21
Lorentz equation, 26
Lorentz force, 67
Lorentz force equation, 8, 38
Lorentz transformation, 15
Lorentz transformations, 37–45
oscillator model, 106, 137–142
Lorentz-Lorenz equation, 131
Lorenz, Ludvig, 42
Lorenz gauge, 42

magnetic constant μ_0, 26
magnetic flux, 2
magnetic monopoles, absence of, 28
magnetization **M**, 109, 125–126
matrix mechanics, 155–157
Maxwell, James Clerk, 1, 6, 29, 39, 40,
52, 61, 149
covariance of Maxwell's equations,
62
Maxwell's equations, 2, 4, 25, 33,
41, 43, 57, 79, 109, 113, 124
Maxwell's equations as first presented in 1864, 7–8
Maxwell's equations in vacuum, 26
Maxwell's stress tensor, 66
theory based on experimental results of Faraday, 4
molecular field, 129
molecular polarizability, 128–129, 138
multipole expansion, 98, 122

Newton, Isaac, 5, 111
Opticks, 149–152
Principia, 5

corpuscular theory, 149
Noether, Emmy
Noether current, 60
Noether's theorem, 58–60

Ørsted, Hans Christian, 2, 28
oersted, unit of magnetic induction, 2
Ohm, Georg Simon
Ohm's law, 26
optical theorem, 145–148

permeability, 109
permittivity, 109
phase velocity, 140
Planck, Max, 69, 137, 153–154
plasma frequency, 140
Poisson, Siméon Denis
Poisson equation, 82
polarization of electromagnetic waves,
34
polarization **P**, 109
potential four-vector, 39, *see* Liénard-
Wiechert potentials
from moving charge, 88
Poynting, John Henry
Poynting vector, 62, 66, 72
Priestly, Joseph, 1

quantum electrodynamics (QED), 4, 157–
159

radiation from an antenna, 106
ray optics, 113
Rayleigh (John William Strutt, 3rd Baron
Rayleigh)
Rayleigh scattering, 128
Rayleigh's law, 123
reflection, 113
reflection coefficient, 115
refraction, 111
refractive index, 110, 128–133, 139–141
relativity, general theory, 4, 19
relativity, special theory, 4, 13, 35
resonance freqency, 138–140
resonant absorption, 140
retarded time, 89

scattering, 121

  differential cross-section, 122
  total cross-section, 122
Schrödinger, Erwin, 155–157
Schwinger, Julian, 157–158
skin depth, 141
Snell's law, 111, 114
standard model, 4, 46, 58, 158–159
Stokes's theorem, 13
stress tensor
  canonical, 61
  canonical stress tensor, 62
  symmetrical stress tensor, 63
structure factor, 123
synchrotron radiation, 104

tensors, 17–19, *see* stress tensor
  Einstein tensor, 19
tesla
  unit of magnetic flux density, 2
  definition of, 31
Tesla, Nikola, 2
time dilation, 19
Tomonaga, Sin-Itiro, 157–158
transmission coefficient, 115

velocity field, 91
volt
  definition of, 31
  unit of electric potential difference,
    2
Volta, Alessandro, 2, 28

wave mechanics, 156
weber, unit of magnetic flux, 2
Weber, Wilhelm, 2

Young, Thomas, 152–153